Lecture Notes in Business Information Processing 495

Series Editors

Wil van der Aalst⬤, *RWTH Aachen University, Aachen, Germany*
Sudha Ram⬤, *University of Arizona, Tucson, AZ, USA*
Michael Rosemann⬤, *Queensland University of Technology, Brisbane, QLD, Australia*
Clemens Szyperski, *Microsoft Research, Redmond, WA, USA*
Giancarlo Guizzardi⬤, *University of Twente, Enschede, The Netherlands*

LNBIP reports state-of-the-art results in areas related to business information systems and industrial application software development – timely, at a high level, and in both printed and electronic form.

The type of material published includes

- Proceedings (published in time for the respective event)
- Postproceedings (consisting of thoroughly revised and/or extended final papers)
- Other edited monographs (such as, for example, project reports or invited volumes)
- Tutorials (coherently integrated collections of lectures given at advanced courses, seminars, schools, etc.)
- Award-winning or exceptional theses

LNBIP is abstracted/indexed in DBLP, EI and Scopus. LNBIP volumes are also submitted for the inclusion in ISI Proceedings.

Jacek Maślankowski · Bartosz Marcinkowski ·
Paulo Rupino da Cunha

Editors

Digital Transformation

15th PLAIS EuroSymposium
on Digital Transformation, PLAIS EuroSymposium 2023
Sopot, Poland, September 28, 2023
Proceedings

 Springer

Editors
Jacek Maślankowski (ORCID)
University of Gdansk
Sopot, Poland

Bartosz Marcinkowski (ORCID)
University of Gdansk
Sopot, Poland

Paulo Rupino da Cunha (ORCID)
University of Coimbra
Coimbra, Portugal

ISSN 1865-1348   ISSN 1865-1356 (electronic)
Lecture Notes in Business Information Processing
ISBN 978-3-031-43589-8   ISBN 978-3-031-43590-4 (eBook)
https://doi.org/10.1007/978-3-031-43590-4

This Springer imprint is published by the registered company Springer Nature Switzerland AG
The registered company address is: Gewerbestrasse 11, 6330 Cham, Switzerland

Paper in this product is recyclable.

Preface

The academic event PLAIS EuroSymposium 2023 was organized with a leading topic of digital transformation. The papers included in the proceedings are related to the use of machine learning, big data and the internet of things in various applications. Other topics of the proceedings concern the current situation of ICT employees and their creativity via social media channels.

The objective of PLAIS EuroSymposium 2023 was to discuss the general issues of digital transformation related topics. The EuroSymposia were initiated by Keng Siau who started this European initiative. Previous EuroSymposia were organized by different academic institutions: University of Galway, Ireland, 2006; University of Gdańsk, Poland, 2007; University of Marburg, Germany, 2008; University of Gdańsk, Poland, 2011–2023.

The accepted papers of previous Gdańsk EuroSymposia were published in the following proceedings:

- 2nd EuroSymposium 2007: A. Bajaj, S. Wrycza (eds), Systems Analysis and Design for Advanced Modeling Methods: Best Practices, Information Science Reference, IGI Global, Hershey, New York, 2009
- 4th EuroSymposium 2011: S. Wrycza (ed.) 2011, Research in Systems Analysis and Design: Models and Methods, series: LNBIP 93, Springer 2011
- Joint Working Conferences EMMSAD/EuroSymposium 2012 held at CAiSE 2012: I. Bider, T. Halpin, J. Krogstie, S. Nurcan, E. Proper, R. Schmidt, P. Soffer, S. Wrycza (eds.) 2012, Enterprise, Business-Process and Information Systems Modeling, series: LNBIP 113, Springer, 2012
- 6th SIGSAND/PLAIS EuroSymposium 2013: S. Wrycza (ed.), Information Systems: Development, Learning, Security, Series: Lecture Notes in Business Information Processing 161, Springer, 2013
- 7th SIGSAND/PLAIS EuroSymposium 2014: S. Wrycza (ed.), Information Systems: Education, Applications, Research, Series: Lecture Notes in Business Information Processing 193, Springer, 2014
- 8th SIGSAND/PLAIS EuroSymposium 2015: S. Wrycza (ed.), Information Systems: Development, Applications, Education, Series: Lecture Notes in Business Information Processing 232, Springer, 2015
- 9th SIGSAND/PLAIS EuroSymposium 2016: S. Wrycza (ed.), Information Systems: Development, Research, Applications, Education, Series: Lecture Notes in Business Information Processing 264, Springer, 2016
- 10th Jubilee SIGSAND/PLAIS EuroSymposium 2017: S. Wrycza, J. Maślankowski (eds), Information Systems: Development, Research, Applications, Education, Series: Lecture Notes in Business Information Processing 300, Springer, 2017
- 11th SIGSAND/PLAIS EuroSymposium 2018: S. Wrycza, J. Maślankowski (eds), Information Systems: Research, Development, Applications, Education, Series: Lecture Notes in Business Information Processing 333, Springer, 2018

- 12th SIGSAND/PLAIS EuroSymposium 2019: S. Wrycza, J. Maślankowski (eds), Information Systems: Research, Development, Applications, Education, Series: Lecture Notes in Business Information Processing 359, Springer, 2019
- 13th SIGSAND/PLAIS EuroSymposium 2021: S. Wrycza, J. Maślankowski (eds), Digital Transformation, Series: Lecture Notes in Business Information Processing 429, Springer, 2021
- 14th SIGSAND/PLAIS EuroSymposium 2022: J. Maślankowski, B. Marcinkowski, P. Rupino da Cunha (eds), Digital Transformation, Series: Lecture Notes in Business Information Processing 465, Springer, 2022

There were two organizers of the 15th EuroSymposium:

- PLAIS – Polish Chapter of AIS
- Department of Business Informatics, University of Gdańsk, Poland

The papers submission and reviewing were evaluated using the conference management system EquinOCS, hosted by Springer. According to the reviews ranking list, 7 papers were accepted for publication in Springer LNBIP volume 495, giving an acceptance rate of 32%.

Topics covered at the symposium included Artificial Intelligence; Big Data; Blockchain Technology; Case Studies; Cloud Computing; Cognitive Issues; Conceptual Modeling; Crowdsourcing and Crowdfunding Models; Data Lakes; Deep Learning; Digital Education; Digital Financial Technologies; Digital Science; Digital Services and Social Media; Education (Curricula, E-learning, MOOCs and Teaching Cases); Emotional Analysis; Entrepreneurial Research; Enterprise Social Networks; Human-Computer Interaction; Industry 4.0; Intelligent Systems; Internet of Things; IS Education and Research (during COVID-19 pandemic); Machine Learning; Mobile Applications; Ontological Foundations; Project Management in Digital Applications; Sentiment Analysis; Social Media Use and Analytics; Social Networking Services; Teams and Teamwork in IS; Text Mining and Web Mining; User Experience (UX) Design; Virtual Reality; Web Intelligence.

We would like to thank all the authors, reviewers, program committee and organizing committee members for contributing to a high-level discussion on the topics of the conference. With their support, the conference EuroSymposium 2023 was a successful event.

September 2023

Jacek Maślankowski
Bartosz Marcinkowski
Paulo Rupino da Cunha

Organization

General Co-chairs

Jacek Maślankowski   University of Gdańsk, Poland
Bartosz Marcinkowski   University of Gdańsk, Poland
Paulo Rupino da Cunha   University of Coimbra, Portugal

Organizers

Polish Chapter of Association for Information Systems (PLAIS)
Department of Business Informatics, University of Gdańsk, Poland

Patronage

European Research Center for Information Systems (ERCIS)
Committee on Informatics of the Polish Academy of Sciences

International Program Committee

Helena Dudycz	Wroclaw University of Economics, Poland
Krzysztof Goczyla	Gdansk University of Technology, Poland
Arkadiusz Januszewski	University of Science and Technology in Bydgoszcz, Poland
Piotr Jędrzejowicz	Gdynia Maritime University, Poland
Dorota Jelonek	Czestochowa University of Technology, Poland
Marite Kirikova	Riga Technical University, Latvia
Vitaliy Kobets	Kherson State University, Ukraine
Jolanta Kowal	University of Wroclaw, Poland
Tim A. Majchrzak	University of Agder, Norway
Mieczyslaw L. Owoc	Wroclaw University of Economics, Poland
Thomas Schuster	Hochschule Pforzheim, Germany
Janice C. Sipior	Villanova University, USA
Catalin Vrabie	National University, Romania
Samuel Fosso Wamba	Toulouse Business School, France
Janusz Wielki	Technical University of Opole, Poland

Andrew Zaliwski   Whitireia Polytechnic Auckland, New Zealand
Iryna Zolotaryova   Kharkiv National University of Economics,
                                        Ukraine

Organizing Committee

Anna Węsierska (Secretary)   University of Gdańsk, Poland
Dorota Buchnowska   University of Gdańsk, Poland
Bartłomiej Gawin   University of Gdańsk, Poland
Przemysław Jatkiewicz   University of Gdańsk, Poland
Dariusz Kralewski   University of Gdańsk, Poland
Jacek Maślankowski   University of Gdańsk, Poland
Patrycja Krauze-Maślankowska   University of Gdańsk, Poland
Michał Kuciapski   University of Gdańsk, Poland
Bartosz Marcinkowski   University of Gdańsk, Poland
Monika Woźniak   University of Gdańsk, Poland

Contents

Using Large Language Models
for the Enforcement of Consumer Rights
in Germany

Lukas Waidelich[1]($\boxtimes$)(ID), Marian Lambert[2](ID), Zina Al-Washash[1](ID),
Steffen Kroschwald[1](ID), Thomas Schuster[1](ID), and Nico Döring[2](ID)

[1] Pforzheim University, Tiefenbronner Str. 65, 75175 Pforzheim, Germany
`lukas.waidelich@hs-pforzheim.de`
[2] XPACE GmbH, Blücherstr. 32, 75177 Pforzheim, Germany

Abstract. In European competition law, consumer protection agencies
and competition authorities play a crucial role in ensuring fair competition. When a violation is identified by these institutions, they typically obtain a cease-and-desist declaration to ensure compliance with
applicable laws. However, the manual verification of compliance is a
time-consuming task, which poses a risk of companies continuing to
engage in unlawful practices to the detriment of consumers. We propose
a technology-enhanced solution to address this issue. Artificial Intelligence emerges as a transformative solution and Large Language Models
now provide the potential for automation, replacing the need for manual
completion of such tedious compliance checks. In our project *KIVEDU*,
we aim to design an AI-based system that automates the enforcement
of consumer rights. In this article, we present an overview of the current state of research, the planned project, the challenges we expect to
encounter, and our initial results as well as planned next steps. With this
work, our goal is to contribute to the enforcement of European consumer
protection law, foster fair competition, and strengthen consumer rights.

Keywords: Consumer Protection · Consumer Rights Enforcement ·
Cease-and-Desist Declaration · Large Language Model · Legal Tech

1 Introduction

In recent years, an exponential increase in the development and online advertise-
ment of a wide range of products has been observed. This trend also led to an
upsurge in consumer-rights infringing behaviors. These behaviors, for example
misleading or false product descriptions and advertising claims, pose a severe
detriment to the rights of consumers and fair market competition.

A 2014 report indicated that approximately 37% of the European Union's
internet commerce did not adhere to Union consumer law, incurring an estimated annual financial damage of €770 million to consumers [12]. It falls to
consumers themselves, as well as consumer protection institutions including consumer protection centers and central competition centers, to identify these violations. Following the identification, a legally binding cease-and-desist declaration

J. Maślankowski et al. (Eds.): PLAIS EuroSymposium 2023, LNBIP 495, pp. 1–15, 2023.
https://doi.org/10.1007/978-3-031-43590-4_1

is obtained - a contract where the offending company pledges to halt the infringing behavior, promising to prevent such infringements in the future.

Enforcing efficient consumer protection delivers multifaceted benefits, not only protecting consumers, but also benefiting law-abiding market participants by fostering fairer competition. Recognizing these beneficial outcomes, consumer rights enforcement has emerged as a key priority for the European Union, which seeks to constrain companies that place their own interests over established laws.

A significant challenge, however, lies in the verification process to ensure that the company abstains from further infringing behavior. This verification is currently a manual process, characterized by its high cost, labor intensity, and time consumption. Consequently, compliance is often not checked at all, potentially allowing companies to continue engaging in consumer-rights infringing behavior.

In the wake of these challenges, the promise held by Artificial Intelligence (AI), particularly by Large Language Models (LLMs), is considerable. AI has the potential to significantly contribute to the enforcement of European consumer rights by automating the process of identifying consumer rights violations. An AI-supported system could enable automated monitoring of cease-and-desist declarations and injunction judgments, notifying relevant parties of potential violations as well as creating and archiving suitable evidence.

This AI-enabled system could provide several key advantages, including improved enforcement of fair competition, enhanced consumer protection, and relief for consumer protection centers. The prospect of AI-powered solutions thus introduces a compelling paradigm shift towards automated consumer rights protection and a robust and fair market ecosystem.

In this paper, we address the mentioned problem and propose potential AI-based solutions for the enforcement of consumer rights within the *KIVEDU* project.[1] To enhance clarity, our paper is focused on three research questions (RQ):

RQ1: How should a technical approach for the automated verification of cease-and-desist declarations in European consumer law be designed?

RQ2: What legal and technical challenges need to be considered during the development?

RQ3: Which LLMs are available and are best suited to tackle these challenges? How do they differ from each other?

The remainder of this paper is structured as follows: Sect. 2 provides a short overview of related work, while Sect. 3 provides a more in-depth description of the KIVEDU project including potential use cases and the proposed software architecture. Section 4 discusses general challenges of implementing the proposed system and Sect. 5 presents our preliminary findings regarding suitable LLMs. Lastly, we conclude and outline our next steps in Sect. 6.

---

[1] The project website can be accessed under https://kivedu-projekt.de.

2   Related Work

We conducted a research of related works in the fields of consumer protection and consumer enforcement focusing on Legal Tech applications. This involved a search of relevant literature and projects in order to gain an understanding of the existing knowledge and initiatives in these areas. We were not able to identify any solutions or approaches that directly or indirectly relate to our descriptive problem. However, there are works that incorporate Legal Tech and are specifically focused on consumer rights, which will be presented now.

Applying machine learning (ML) algorithms to consumer protection legal texts has been a focus of research for several years. Initial approaches were primarily directed at the analysis of standard contracts such as Terms of Service (ToS) and privacy agreements. One of the earliest research attempts to identify unfair clauses in online consumer contracts was by Lippi et al. [17]. They used a simple method that treated the contract as a "bag of words" and identified patterns within these words using ML techniques such as Support Vector Machines (SVM) and Tree Kernels. Their work set an important benchmark given that their approach far outperformed a random guessing strategy.

Later, Braun et al. (2019) outlined how Legal Tech can be used to empower consumers in the digital age, presenting two prototypes that semantically analyze, evaluate, and summarize the ToS of German web shops [6]. The same authors also focused on the automatic detection of illegal clauses in German Terms and Conditions (TaC) [5]. They used a pre-trained German language model (BERT [11]), which was successful in identifying illegal texts with a very high accuracy. Drawing on these findings, the CLAUDETTE tool was developed, which in addition to expanding the underlying dataset to 50 ToS, also applies other ML systems (including deep learning) to carry out more granular classification into individual unfairness categories [18]. This tool has also been used to identify unclear or unfair clauses in privacy statements with regard to the European General Data Protection Regulation (EU-GDPR) [9].

Apart from detecting unfair or illegal clauses in online standard contracts, ML has also been applied to the analysis and identification of judgment documents of US trademark disputes as precedent cases by Trappey et al. [29] and in the domain of patent law to predict legal disputes [13,19].

Furthermore, Chakrabarti et al. (2018) developed a system based on ML for the identification and extraction of risky paragraphs in contracts and their subsequent classification into risk classes [7]. The authors employed paragraph vectors in training and utilized classification algorithms that included various variations of SVM and Naive Bayes models, achieving high performance scores.

Our project differs from these publications in several aspects. Firstly, it does not deal with standard contracts such as ToS or privacy statements. Instead, it focuses on cease-and-desist declarations, which are civil law contracts between companies and consumer protection centers or competition associations. These contracts tend to be highly individualized, posing challenges for ML training and processing. Additionally, a legal violation does not solely rely on the text of the cease-and-desist declaration but also depends on the corresponding online content which might contain the violation. Consequently, it is not possible to mark entire clauses as unfair, unlike the approach taken in the mentioned publications. Instead, a thorough individual examination of the cease-and-desist declaration, taking into account the specific text content and corresponding context, is required.

3　AI-Based Detection of Cease-and-Desist Violations

In this section, a summary of the key elements of the proposed project is presented, including the concept, the high-level software architecture, and first potential use cases. Our proposal focuses on automating the examination of cease-and-desist declarations in compliance with European consumer law though we acknowledge that there might a wide range of other use cases (see Sect. 3.3).

3.1　Concept

In the pursuit of safeguarding consumer rights, fostering fair competition, and ensuring regulatory compliance, this project aims to develop an AI-driven software system specifically designed to identify and record cease-and-desist declaration violations on the Internet. The envisioned system operates in the background, conducting regular scans across the web to pinpoint potentially infringing websites for which a cease-and-desist declaration is available. It then scans these websites and applies AI models to search violations of cease-and-desist declarations. Once a violation is detected, the system notifies all parties involved. Our interdisciplinary project team, which comprises both technical experts specializing in AI and software development, and legal professionals, is working collaboratively to tackle the inherent technological and legal challenges. Moreover, our partnership with a consumer protection center offers us the advantage of a substantial training dataset for enhancing the AI's ability to recognize claims and perform compliance checks. Presently, the project is in its nascent stage, with ongoing research efforts being devoted to laying the groundwork for subsequent phases. This paper intends to outline the current progress and the innovative potential of this automated, AI-driven violation detection and reporting system.

3.2　System Architecture

The software architecture of the system (depicted in Fig. 1) is primarily designed around the principles of service-oriented architecture (SOA), event-driven and loosely-coupled design, and employs a variety of cloud-native components to handle specific tasks such as data storage, text extraction and data analysis. As a SOA, it promotes reusability and modularity of components, fostering efficient development and maintenance by dividing the system into well-defined services. Its event-driven nature enables the system to be highly responsive and scalable, as components are triggered to perform their tasks based on specific events, allowing for on-demand resource usage. Lastly, the loosely-coupled design provides flexibility and robustness, as changes to one component do not necessitate changes in others. This allows for easier upgrades, modifications, and debugging, contributing to overall system resilience and evolution.

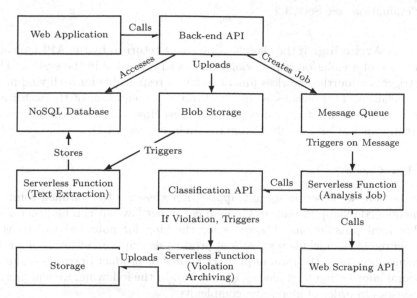

Fig. 1: Proposed software architecture for the project

Users interact with the system through a web application. This application is designed to be the primary point of user-system interaction, providing functionalities such as uploading PDF files of cease and desist declarations and creating new analysis jobs. The web application integrates with a back-end API, which provides the necessary functionality for data manipulation and retrieval. This API communicates with a NoSQL database, storing and fetching data as required. The use of a NoSQL database enables flexibility in the data model, catering to diverse data types and structures inherent in the project.

PDF Upload. When a user uploads a new cease and desist declaration, the file is sent to the API, which then stores it in a blob storage container. The upload event triggers a serverless function that identifies and extracts all text from the PDF and stores this extracted text on the database item associated with the cease and desist declaration.

Analysis Job. When a user initiates a new analysis job via the web application, another serverless function is triggered which in turn pushes a new message onto a queue. The creation of the message triggers a third serverless function, which runs the analysis job by making calls to a semantic web scraping and classification API. The web scraping API is responsible for extracting relevant information from the website which should be examined. The classification API takes the cease and desist declaration text and extracted data as input and returns a classification result indicating whether a violation is present. For this, it employs a large language model. Which model will be used is currently still under evaluation (see Sect. 4.2).

Violation Archiving. If the classification result returned by the API indicates the presence of a violation, a new violation record is created in the system. This event triggers a fourth serverless function that is responsible for archiving proof of the violation. This ensures a time-accurate representation of the website is preserved in the event of any subsequent changes. How this proof will be created is currently unclear and requires more research (see Sect. 4.2).

3.3   Use Cases

In this section, we describe specific application areas for the outlined solution. The mentioned examples refer to German consumer law but can be generalized to other legal jurisdictions. The basis for checking for potential violations of consumer rights can include text, such as product descriptions or terms of service documents, images in the form of product photos or product drawings and even behavioral information (e.g. placing of cookies). In the following, we will present the use cases in order of increasing complexity.

Food Labeling and Other Information. Food labeling is heavily regulated by various authorities worldwide to safeguard consumers from being misled by false or exaggerated information. For instance, descriptors like "wholesome" are not permitted to be used in relation to certain products, such as coffee. Yet, the monumental task of manually examining each product's description for violations proves to be arduous, if not impossible, particularly with the large amount of products available online. This not only applies to food labeling but also other misleading product specifications. Our system could prove useful by automatically identifying and archiving such violations. This is the use case we put our primary focus on.

Compliance with legal (informational) Obligations. Compliance with legal obligations is critical during the purchasing process, as they serve to inform consumers about their rights, fostering an environment of transparency and trust. For instance, the law expressly prohibits distance selling without sufficient information provided to consumers about the withdrawal process, including the provision of an official withdrawal form. Failure to comply with these obligations can lead to consumers making ill-informed decisions and consequently impairing trust in the purchasing process. To mitigate these risks, our system could potentially automate the process of verification, ensuring that all legal obligations are consistently and accurately met, thus safeguarding consumer rights and maintaining the integrity of the transaction process.

Data Protection. In today's digital era, the protection of consumer data, especially online, has become paramount. Proper data protection safeguards the privacy of consumer data, prevents misuse of information, and fosters a sense of trust and transparency between consumers and businesses. Moreover, it is crucial for consumers to have control over what data they choose to share and to be well-informed about how this data is being used. For instance, websites are legally obliged not to set analytical cookies until they have obtained explicit consent from consumers. In such a context, our system might play a vital role by automatically monitoring and ensuring compliance with various data protection obligations, thereby safeguarding consumers' data privacy rights.

Trademark Usage and Design Patents. The infringement of trademark rights is not only prohibited but also legally punishable, as such violations can breed unfair competition and cause harm to both businesses and consumers. Additionally, these infringements often result in the dissemination of misleading information, which can result in consumers making uninformed and misguided decisions. A specific example of this can be seen with online platforms, where it is illegal to announce, offer, or sell items that were produced and/or first placed on the market without the express permission of the rights holder, particularly if these items are protected under trademark law. Similarly, protected design patterns may not be used or imitated without the consent of the rights holders. As a specific example, it is prohibited to use a protected car key case in the EU, particularly by introducing, offering, promoting, or distributing it there, or by possessing it for these purposes, if it has not been placed on the market with the consent of the rights holder. To counter such activities, our system can potentially detect trademark or patent infringements, thus ensuring compliance and mitigating potential harms associated with these transgressions.

4 Challenges

The development, the provision and operation, hence the product life cycle of the software system aimed for in our project raises legal questions that need to be answered in the course of our project and involve various and technical difficulties. These problems can arise due to the complex legal requirements and intricate technical aspects of the development process. In the context of our proposed approach, we have identified and categorized these challenges into legal and technological facets.

4.1 Legal

General. When it comes to the creation and application of a new system, the legal landscape is not just a background consideration - it forms the very foundation upon which developers, providers, and users interact. For a system to be marketable and operational in the long run, it is crucial to weave legal considerations into its very design, including the product's business model, processes, and the user journey.

Legal requirements often directly impact the product and its delivery. Non-compliance can lead to prohibitions from authorities, reluctance from potential marketers, or even legal liabilities. However, it is also essential to remember that users, as potential customers, are bound by their own set of legal requirements. In creating a product that enables and supports users in abiding by the law, we build trust and increase our marketability.

As a result, our approach focuses on "by-design" life cycle compliance. It is imperative that we systematically identify and understand the legal requirements in various stages of the product's life and how they apply to different stakeholder groups. Both the software product architecture and the business models contain elements that could trigger different legal requirements.

While we acknowledge that operating in the digital space may expose us to international jurisdictions, our primary attention lies on the legislation within Europe and Germany. When considering the technological scope, the usage objectives, architectural design, potential supply chains, and the prospective software market, we have pinpointed three main areas of law to focus on. These are the General Data Protection Regulation "GDPR", the Artificial Intelligence Act "AI-Act" and the intellectual property law.

Given the software's primary functionality will be the enforcement of consumer law, it is necessary to define the functional requirements for use cases to which the software will be applied. To do this, we need to identify cases relevant to consumer law-related cease-and-desist declarations. An initial roadmap for potential legal applications was outlined in Sect. 3.3. However, we are aware that new use cases may be added as the project progresses, especially taking into account new court rulings.

The following is a general outline of the legal issues that have arisen so far in connection with the project. These issues will be further evaluated and continuously adapted in the course of the project.

EU-GDPR. The architecture outlined above is mainly based on the processing of data. As a consequence, the laws regulating the handling of data are the main focus of our legal assessment. Retrieving information from declarations and court decisions as well as crawling through databases and websites creates large amounts of data. Applying language models as described above leads to the processing of such data. Data legislation of the EU might be affected. One of the important areas of investigation from a legal perspective is certainly the EU-GDPR. As soon as this data is directly or indirectly related to a natural person, it is *personal data* within the meaning of the EU-GDPR (Art. 4 I EU-GDPR). The collection, retrieving or combination of this information hence constitutes a processing of personal data (Art. 4 II EU-GDPR) and the scope of the EU-GDPR is opened (Art. 2 I EU-GDPR). Among others, EU-GDPR requires that personal data is processed lawfully and fair (Art. 5 Ia EU-GDPR). This raises the question of the permissibility of the processing (Art. 6 I EU-GDPR) in the respective elements of the software and user journey. In addition, further principles for the processing of personal data, such as the principle of transparency and purpose limitation (Art. 5 Ia and b EU-GDPR), must be observed. Personal data processed needs to be limited to specified, explicit and legitimate purposes and to an extent necesssary for these purposes (Art. 5 Ib EU-GDPR). Furthermore, the rights of the data subjects must be taken into account (Art. 12-23 EU-GDPR). For example, the data subject has a right to be forgotten (Art. 17 EU-GDPR) and there should an option to erase one's personal data. Such deletion orders could become problematic if, when using an external LLM service, control over the data fed in no longer lies with the user/developer. The principle of "privacy by design" must also be taken into account as early as the development phase. To guarantee the conformity with the law, such requirements need to be taken into account and measures have to be implemented when determining the means of processing according to Art. 25 of the EU-GDPR [24, 30].

When data is being processed outside of the EU - e.g. due to service providers commissioned or (cloud) services made use of - specific requirements and limitations for the transfer of personal data have to be considered (Art. 44-50 EU-GDPR). When using open source applications with third-country references (all countries that do not belong to the European Economic Area), the legal risks and framework conditions of the third-country transfer are of importance.

In the EU-GDPR part of the project, it is important to clearly define data protection requirements, to take them into account during development and to identify possible data protection risks. Since EU-GDPR follows a risk based approach it will be necessary to systematically categorize risks when assessing the legal requirements and developing design patterns and recommendations for the design on a basis that allows to scale the product and its elements alongside the different parameters of risk.

EU AI-Act. In addition to data protection, the legal developments with regard to the regulation of AI systems will be observed and taken into account within the framework of the project. The EU Commission is currently working on a regulation (AI-Act) at the EU level to regulate the use of AI. The AI-Act will be relevant for developers, providers, users and operators of AI systems, among others. The aim is to create a legal framework for the development, deployment and use of AI systems. Among other things, it is intended to ensure that AI systems are transparent, reliable and secure. In addition, minimum requirements are defined in order to meet the aforementioned principles of traceability, reliability and security. In light of this, the development of the AI Act is of high importance for the software design [15].

Intellectual Property. Intellectual property law also plays an important role. When developing artificially intelligent software, developers should understand their intellectual property rights and take appropriate measures to protect them. This includes, for example, the use of non-disclosure agreements. In addition, when using open source applications, the rights of the providers must be respected in order to avoid copyright or licence infringements.

4.2   Technical

LLM Selection. The task at hand - detecting violations of cease-and-desist declarations in online content - necessitates the employed AI model to have a high level of language understanding, robust reasoning capabilities, and adept logical inference. To effectively carry out this task, the model must not only comprehend the complexities of the legal language embedded in cease-and-desist declarations but also discern any potential violations in the context of a diverse range of online content. Training a model from scratch that can accurately exhibit these capabilities is an enormous challenge. It demands substantial computational resources, considerable time investment, and extensive training data. Moreover, constructing a model that can comprehend language at this level also requires deep expertise in natural language processing and machine learning.

In order to circumvent these obstacles, we plan to leverage existing LLMs that have been pre-trained on massive text corpora. These LLMs have been trained to grasp the intricacies of human language and to generate coherent and contextually relevant responses. Consequently, they already possess a significant level of understanding and inference capabilities that can be fine-tuned for the specific task of detecting cease-and-desist declaration violations.

However, the selection of an appropriate LLM is a nuanced process. These models can vary considerably in several aspects, most notably the performance (i.e. how well they perform at detecting violations), processing speed, cost, licensing model and data privacy guarantees they provide. All of these elements can significantly impact the model's suitability for our project.

Given the critical role of the LLM in detecting cease-and-desist violations, the model selection process is of great importance. A suboptimal choice can significantly reduce the effectiveness of our system and may lead to erroneous violation detection. Therefore, we plan to conduct an exhaustive research of potential LLMs, weighing their relative strengths and weaknesses for the task at hand. A preliminary research was already carried out and is presented in Sect. 5. We will apply each potential LLM to a curated set of test cases that capture the complexity and diversity of cease-and-desist violations in the online environment. This empirical approach will facilitate the selection of the most suitable LLM by directly comparing their performances, thereby ensuring the highest level of accuracy and efficiency in violation detection.

AI Explainability. Our system operates at a juncture of high importance; the decisions made are consequential and potentially bear significant impact. However, a key challenge lies in the inherent complexity of these decisions. Given the multiple facets of language understanding and logical inference involved in identifying cease-and-desist violations, the reasoning behind the outcome might be challenging for users to fully comprehend.

This complexity underpins a broader issue facing artificial intelligence (AI) systems - the challenge of interpretability and explainability. Given that these systems often rely on intricate and high-dimensional data representations, understanding the reasoning behind their decisions can be quite challenging. This issue is very pronounced in the field of large language models, and is currently an active area of research [14].

In our system, we are cognizant of this challenge and strive to enhance the explainability of AI decisions. Our aim is to render the AI decision-making process comprehensible, offering detailed explanations for each decision. We believe that users should not just be presented with an outcome, but also understand the reasoning that led to it.

Violation Evidence Archival. One of the significant challenges in handling violations of cease-and-desist declarations lies not only in their identification but also in the subsequent archiving and securing of credible evidence that can withstand legal scrutiny. This is particularly challenging due to the fluidity of online content, the ambiguity of what constitutes as proof, and the potential for digital manipulation.

Firstly, it is currently unclear what exactly can serve as valid evidence in the case of a violation. Screenshots or static exports of the offending website might be viable, but these methods have their limitations. Screenshots, for instance, can easily be manipulated, and the dynamic content of websites might not be fully captured in a static export.

Additionally, preserving the integrity of evidence is a challenge. It is essential to store the proof in a manner that precludes any form of deletion or manipulation. Traditional methods of digital data storage are susceptible to tampering and unauthorized modifications, raising concerns about the validity and reliability of the archived proof. Blockchain technology could potentially be a solution to these challenges as it maintains a growing list of records (blocks) which are resistant to modification once validated but this stands to be examined further.

5   Current State of Large Language Models

As a crucial component of this project, an LLM is responsible for comparing the content of the cease-and-desist declaration with the examined contents (e.g., product descriptions) and identifying potential violations. In addition to well-known LLMs like ChatGPT, there are a number of other LLMs that differ in their characteristics and capabilities. As described before, identifying the optimal LLM with regards to power, performance, cost, data privacy guarantees and other factors will present a significant challenge in the project. During our preliminary research, we have collected and compared a small number of LLMs which might be suitable for application in this project. Table 1 offers a summary of these LLMs. We have focused on a manually chosen set of LLMs that we estimate to at least match or surpass the capabilities of GPT-3.

Within the table, each model's specifics are included, including its type, size and context window (described in terms of parameters and tokens) and the associated licensing terms under which the model can be used.

As for the *Type* column, a value of *IT* is indicative of instruction-tuned models, whereas *BM* refers to base models. Instruction-tuned models are pre-trained to adhere to instructions more strictly, typically via a method known as

Reinforcement Learning from Human Feedback (RLHF) [8]. Some models are available as both IT and BM (*IT/BM*) while others provide only an experimental IT version (*BM (IT)*). The *Context Window* column describes the number of tokens the model can use to make predictions. A token represents a small number of continuous characters (typically the length of a syllable) which the language model processes as an atomic text entity. If the context window is too small, the model might not be able to process longer cease-and-desist declarations and domain contents, making it significantly less useful for this project. The *Size* column contains the number of parameters (in million, billion or trillion) the model consists of. Generally speaking, more parameters (i.e. higher model complexity) might allow the model to have a higher level of language understanding, though this relation does not seem to apply reliably (see [31]).

Table 1: Overview of potential suitable LLMs

Model	Organization	Type	Context Window (Tokens)	Size (Parameters)	License
ChatGPT [21]	OpenAI	IT	4097	175B	Proprietary
GPT-4 [22]	OpenAI	IT	8192, 32768	170T	Proprietary
LLaMA [28]	Meta	BM	2048	7B, 13B, 33B, 65B	NC Custom
Alpaca [27]	Stanford	IT	2048	7B	NC Custom/ CC BY-NC-SA 4.0
MPT [20]	MosaicML	IT/BM	2048	7B	Apache-2.0/ CC BY-SA-3.0
OpenAssistant [16]	LAION	IT	Varies	Varies	NC Custom/ CC BY-SA-4.0/ Apache-2.0
StableLM [25]	StabilityAI	BM (IT)	4096	3B, 7B, 15B, 65B	CC BY-SA-4.0/ CC BY-NC-SA 4.0
StableVicuna [26]	StabilityAI	IT	2048	13B	NC Custom/ CC-BY-NC-SA-4.0
Claude [2]	Anthropic	IT	100k	52B	Proprietary
Luminous [1]	Aleph Alpha	BM/IT	unknown	13B, 30B, 70B	Proprietary
PaLM 2 [21]	Google	IT	unknown	unknown	Proprietary
Dolly-v2 [10]	Databricks	IT	2048	3B, 7B, 12B	MIT
Pythia [3]	EleutherAI	BM	2048	70M, 160M, 410M, 1B, 1.4B, 2.8B, 6.9B, 12B	Apache-2.0
BLOOM [4]	BigScience	BM	unknown	176B	RAIL (open)
LIMA [31]	Meta	IT	2048	65B	unknown
Falcon [23]	Technology Innovation Institute	BM/IT	2048	7B, 40B	Apache 2.0

In our preliminary research, we have most notably identified a tradeoff between performance and privacy when selecting the appropriate LLM. Currently, leading proprietary models like ChatGPT demonstrate exceptional performance but come with limited data sovereignty, which may also vary based on applicable data privacy regulations. On the other hand, open source models can be deployed on-premise, ensuring complete data privacy, but often exhibit inferior performance. However, it is worth noting that open source models have made significant strides in performance over the past few months. If this trend continues, they could present a viable alternative to proprietary models, addressing privacy concerns effectively. Based on our initial evaluation, we find the Luminous (proprietary) and MPT (open source) models particularly promising at present as they provide a good balance between performance and data privacy. Nevertheless, the field of LLM research is dynamic, and our ongoing work on this project is still in its early stages, prompting us to closely monitor advancements in this area and update our model selection accordingly.

6   Conclusion and Outlook

To the best of our knowledge, the *KIVEDU* project presented in this paper is the first attempt at utilizing an AI system to automatically monitor, analyze, and document violations of cease-and-desist declarations. We systematically pursue this endeavor from requirements engineering to analysis & design, implementation, and prototypical deployment, with the goal of strengthening consumer rights in the EU and Germany. In this paper, we describe the idea, provide preliminary insights and raise awareness of the project. Following a literature review in the second chapter, we describe the project idea, intended software architecture and potential use cases in Sect. 3, answering **RQ1**.

Software development and AI are both multifaceted fields that bring both legal as well as technical challenges which we describe in Sect. 4 to provide an answer to **RQ2**. Whether it is regulatory requirements regarding data protection and privacy, or recently emerging AI legislation, all legal challenges need to be adequately considered. As for technical challenges, issues such as AI explainability and proper archival of violation proofs are described. In particular, the selection of a suitable LLM is identified as a challenge, as the LLM is the central component of our system that determines whether a violation of consumer rights is present. The market for LLMs is dynamic and currently lacks clarity. By presenting an overview of current proprietary and open source LLMs, we provide insights into existing models and categorize them based on relevant criteria to answer **RQ3**.

In this publication, we present the idea of automated consumer enforcement and aim to inspire further work in this field. In the future, comprehensive requirements engineering, and a specific data protection concept needs to be developed to address the challenges identified. The definition of use cases should also be further explored. For this purpose, collaborations with consumer protection centers, competition associations, and other organizations such as law firms are envisaged. Also, data on consumer violations must be gathered for the purpose of fine-tuning and testing the AI models.

References

1. Aleph Alpha: Luminous (2023). https://www.aleph-alpha.com/luminous
2. Anthropic: Introducing Claude (Mar 2023). https://www.anthropic.com/index/introducing-claude
3. Biderman, S., et al.: Pythia: A Suite for Analyzing Large Language Models Across Training and Scaling (May 2023). https://doi.org/10.48550/arXiv.2304.01373
4. BigScience Workshop: BLOOM: A 176B-Parameter Open-Access Multilingual Language Model (Jun 2023). https://doi.org/10.48550/arXiv.2211.05100
5. Braun, D., Matthes, F.: NLP for Consumer protection: battling illegal clauses in german terms and conditions in online shopping. In: Proceedings of the 1st Workshop on NLP for Positive Impact. Association for Computational Linguistics, Online (Aug 2021). https://doi.org/10.18653/v1/2021.nlp4posimpact-1.10
6. Braun, D., Scepankova, E., Holl, P., Matthes, F.: Consumer protection in the digital era: the potential of customer-centered LegalTech. In: David, K., Geihs, K., Lange, M., Stumme, G. (eds.) INFORMATIK 2019: 50 Jahre Gesellschaft für Informatik - Informatik für Gesellschaft, pp. 407–420. Gesellschaft für Informatik e.V., Bonn (2019). https://doi.org/10.18420/inf2019_58
7. Chakrabarti, D., et al.: Use of artificial intelligence to analyse risk in legal documents for a better decision support. In: TENCON 2018–2018 IEEE Region 10 Conference, pp. 683–688. IEEE, Jeju, Korea (South) (Oct 2018). https://doi.org/10.1109/TENCON.2018.8650382
8. Christiano, P., Leike, J., Brown, T.B., Martic, M., Legg, S., Amodei, D.: Deep reinforcement learning from human preferences (Feb 2023). https://doi.org/10.48550/arXiv.1706.03741
9. Contissa, G., et al.: Claudette meets GDPR: automating the evaluation of privacy policies using. Artif. Intell. (2018). https://doi.org/10.2139/ssrn.3208596
10. Databricks: Free Dolly: Introducing the World's First Truly Open Instruction-Tuned LLM (Apr 2023). https://www.databricks.com/blog/2023/04/12/dolly-first-open-commercially-viable-instruction-tuned-llm
11. Devlin, J., Chang, M.W., Lee, K., Toutanova, K.: BERT: Pre-training of Deep Bidirectional Transformers for Language Understanding (May 2019). https://doi.org/10.48550/arXiv.1810.04805
12. European Commission: 2016/0148 (COD) Cooperation between national authorities responsible for the enforcement of consumer protection laws (May 2016). https://eur-lex.europa.eu/legal-content/EN/TXT/?uri=celex:52016PC0283
13. Juranek, S., Otneim, H.: Using machine learning to predict patent lawsuits (2021). https://doi.org/10.2139/ssrn.3871701
14. Kasneci, E., et al.: ChatGPT for good? on opportunities and challenges of large language models for education. Learn. Individ. Differ. **103**, 102274 (2023). https://doi.org/10.1016/j.lindif.2023.102274
15. Kroschwald, S.: Nutzer-, kontext- und situationsbedingte Vulnerabilität in digitalen Gesellschaften: Schutz, Selbstbestimmung und Teilhabe by Design vor dem Hintergrund des Art. 25 DSGVO und dem KI-Verordnungsentwurf. Zeitschrift für Digitalisierung und Recht (1), 1–22 (2023)
16. Köpf, A., et al.: OpenAssistant Conversations - Democratizing Large Language Model Alignment (Apr 2023). https://doi.org/10.48550/arXiv.2304.07327
17. Lippi, M., et al.: Automated detection of unfair clauses in online consumer contracts. In: Legal Knowledge and Information Systems, pp. 145–154. IOS Press (2017). https://doi.org/10.3233/978-1-61499-838-9-145

18. Lippi, M., et al.: CLAUDETTE: an automated detector of potentially unfair clauses in online terms of service. Artificial Intell. Law **27**(2), 117–139 (2019). https://doi.org/10.1007/s10506-019-09243-2

19. Liu, Q., Wu, H., Ye, Y., Zhao, H., Liu, C., Du, D.: Patent litigation prediction: a convolutional tensor factorization approach. In: International Joint Conference on Artificial Intelligence (2018). https://doi.org/10.24963/ijcai.2018/701

20. MosaicML NLP Team: Introducing MPT-7B: A New Standard for Open-Source, Commercially Usable LLMs (2023). www.mosaicml.com/blog/mpt-7b

21. OpenAI: Introducing ChatGPT (Nov 2022). https://openai.com/blog/chatgpt

22. OpenAI: GPT-4 Technical Report (Mar 2023). https://doi.org/10.48550/arXiv.2303.08774

23. Penedo, G., et al.: The RefinedWeb Dataset for Falcon LLM: Outperforming Curated Corpora with Web Data, and Web Data Only (Jun 2023)

24. Rösch, D., Schuster, T., Waidelich, L., Alpers, S.: Privacy control patterns for compliant application of GDPR. In: AMCIS 2019 Proceedings (Jul 2019). https://aisel.aisnet.org/amcis2019/info_security_privacy/info_security_privacy/27

25. Stability AI: Stability AI Launches the First of its StableLM Suite of Language Models (Apr 2023). https://stability.ai/blog/stability-ai-launches-the-first-of-its-stablelm-suite-of-language-models

26. Stability AI: Stability AI releases StableVicuna, the AI World's First Open Source RLHF LLM Chatbot (Apr 2023). https://stability.ai/blog/stablevicuna-open-source-rlhf-chatbot

27. Taori, R., et al.: Stanford Alpaca: An Instruction-following LLaMA model (2023). https://github.com/tatsu-lab/stanford_alpaca

28. Touvron, H., et al.: LLaMA: Open and Efficient Foundation Language Models (Feb 2023). https://doi.org/10.48550/arXiv.2302.13971

29. Trappey, C.V., Trappey, A.J.C., Liu, B.H.: Identify trademark legal case precedents - Using machine learning to enable semantic analysis of judgments. World Patent Information 62 (Sep 2020). https://doi.org/10.1016/j.wpi.2020.101980

30. Vásquez, S., Kroschwald, S.: Data-driven vehicles: Privacy by Design - Verantwortung zwischen Herstellern und Anbietern und das Principal-Agent-Problem. Zeitschrift für IT-Recht und Recht der Digitalisierung **4**, 217–221 (2020)

31. Zhou, C., et al.: LIMA: Less Is More for Alignment (May 2023). https://doi.org/10.48550/arXiv.2305.11206

Creating of a General Purpose Language for the Construction of Dynamic Reports

Vlad Iatsiuta , Vitaliy Kobets[(✉)] , and Oleksii Ivanov

Kherson State University, 27, Universitetska St., Kherson 73003, Ukraine
{vladyslav.yatsiuta,oleksii.ivanov}@university.kherson.ua,
vkobets@kse.org.ua

Abstract. In digital environment, generating reports is an integral part of any business, and therefore there is a huge need for a tool for creating complex adaptive reports in investment, marketing activities, financial projects etc. The purpose of the paper is to formulate the requirements for GPL for the creation of custom reports in different business areas and build a software product that will use it. We have researched the possibilities of creating a universal language for building complex reports that is flexible enough to be used in any business domain. We have also identified the main requirements for such a language and the software product that would utilize it. We have paid particular attention to the peculiarities and problems associated with the creation and use of such a language, and have proposed ways to address them. As an experiment, we have created a prototype software module using a language based on mathematical formulas. The developed module can be used both for reports and for any calculations of companies engaged in product and business analytics.

Keywords: General-Purpose Language · Domain-Specific Language · Report · Software Module · Calculations

1 Introduction

Building reports is an integral component of any business, from calculating simple statistics to complex mathematical calculations that affect future investment direction. For example, in marketing, reports are actively used to analyze the performance of various advertising campaigns based on data such as the number of views, time spent viewing ads, and bounce rate. Sales-oriented businesses build their strategies on product data, regions, and targeted audiences. In financial projects, analysis of gross profit margin, net profit margin, and ROI (return on investment) is extremely important for increasing revenue and reducing costs. Human resources departments of any company carefully study reports on employee performance to understand the need for additional training programs and the provision of certain types of assistance using KPI. These reports are usually based on criteria such as employee turnover rate, employee satisfaction scores, and performance ratings.

J. Maślankowski et al. (Eds.): PLAIS EuroSymposium 2023, LNBIP 495, pp. 16–37, 2023.
https://doi.org/10.1007/978-3-031-43590-4_2

Each business direction is unique, and even companies operating in the same field with similar input data usually have different approaches to their analysis and management. However, every business shares the need for collecting, transforming, and researching data presented in a concise format such as a report. The research question arises whether modern methodologies and technologies can be used to create a unified way of creating reports that is not dependent on their area of use. To answer this question, a set of requirements for such a tool needs to be formulated first. In turn, to formulate requirements correctly, we need to follow the general standards [1] and there needs to be a clear understanding of what unites different business areas and what distinguishes them from each other. Since this tool will only work with data sets, comparisons need to be made from the standpoint of the data.

One of important skill of business analytics is a report preparation, which gives base for decision-making [2]. Using initial inputs of investors robo-advisor can create their risk profile [3] and generates their investment portfolios as report with proportion distribution of different financial instruments [4, 5]. Search for matches between the competencies of employees and the requirements of employers leads to the generation of a report with candidates for job vacancies in the labor market [6].

Also, for the full functionality of this tool, it is necessary to create a universal language GPL (general purpose language) for generating report templates, which should contain elements that allow it to be adaptable to the needs of any business. The **purpose** of the paper is to formulate the requirements for GPL for the creation of custom reports in different business areas and build a software product that will use it.

As an experiment, we use the formulated GPL to create a software application for calculating business metrics and subsequently including them in the report. For this purpose, it is not enough to just formulate the GPL, additionally, we need to create a mechanism for its processing. In this paper, we will explore the possibilities of such processing using existing systems as well as algorithms for creating such systems independently.

The paper is structured as follows: Sect. 2 considers related works, Sect. 3 frames methodological approach and requirements; Sect. 4 presents the results; Sect. 5 proposes experiments with program module execution; the last section contains the conclusions.

2 Related Works

There are many studies exploring the possibilities of data processing and creating complex reports [7]. 'Such techniques typically target the Data Access Layer of a tool to access and manipulate its content, with the assumption that sufficient information and control is available within this layer to automate the process' [8]. Software tools will most likely adopt different interface technologies, making their interoperability difficult. Investigators points out that integration compensates the deficiencies in existing tools and technologies. Many software tools have no support of write-access to the exposed resources [8]. A tool interoperability approach based on the Linked Data principles, which can mitigate the identified shortcomings. Lu [9] describes the syntactic complexities of language creation, including report-building languages [10], and also raising issues of data quality [11]. 'Seamless data communication in all phases of the product development process is a prerequisite for cost-optimal and successful collaboration processes, but the main problem in this process chain is the insufficient data quality' [11].

Information extraction is important for data transformation in business cases of digital economics. However, building extraction systems in real cases has two challenges: (i) the availability of labeled data as a rule is bounded and (ii) classification is required in detail [12]. Information extractor can use pre-trained language models trained by transformers for the extraction, outputs are stored in a database for document processing (Fig. 1).

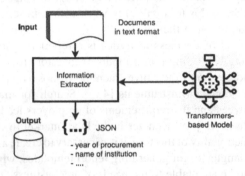

Fig. 1. The overview of the system [12]

The task of the system is to convert unstructured documents to structured data [12]. Business valuation contains collecting, analysis, and applying to financial integral information to estimate the business value. The evaluated results are used in the decision making of pricing, market segmentation, business capacity, cost per unit etc. 'There are specific information and events about business valuation stored in the Intelligent financial statements presented by the HTML and PDF files. Hence, there are demand on information extraction system of financial data for business valuation from the domestic business financial statements from different heterogeneous data sources' [13].

'The application of Business Intelligence systems can be seen as a business strategy and development, which integrates a comprehensive set of services to provide relevant corporate information in strategic and operational decision-making, and to increase the corporation's competitiveness. For the successful implementation of a BI system in an organization, it is necessary to use well-defined processes and business rules' [14]. 'Computer applications have allowed organizations to have quality control over data, generate relevant indicators to be provided to managers about the organizations business, preparing them to design forecast scenarios more effectively' using different technological arcitecture (Fig. 2) [14].

There are many domain-specific language (DSL) for different but not for general purposes. For example, domain-specific language (DSL) can describe machine learning datasets in terms of their structure, background, and social concerns (fairness, absence of bias) which can facilitate selecting of the most appropriate dataset and its usability for a new project or startups [15]. Quintero et al. [16] propose a 'Domain-Specific Language to facilitate the modeling of Usage Control model policies and their integration in model-based development processes' to facilitate security policy.

Research [17] propose a Domain-Specific Language (DSL) based on Set Theory for requirement analysts, where simple visual language can automatically specify software requirement structural invariants and can disclosure requirement inconsistencies.

Due to DSL level of abstraction, they enable building applications that simplify software implementation. Developed architecture of programming model [18] allows the definition and specification of JSON-DSLs, the implementation of JavaScript components, the use of connectors to heterogeneous information sources, and the integration with web components and JavaScript frameworks at both the web server and web client level. The hierarchies of Multilevel Modelling approach may represent domain specific modelling languages that can co-exist and evolve independently and simultaneously in addition to their participation in compositions [19].

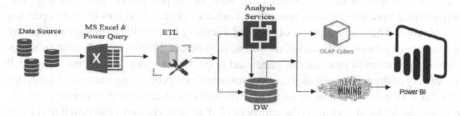

Fig. 2. Technical architecture [14]

Knowledge captured in process models must be consistently transferred to the specified requirements; for business-related software systems final requirements are typically still specified in natural language, so problem of inconsistency appears between process models and natural language requirements in the context of software development [20]. To address this problem, Aysolmaz et al. propose a semi-automated approach whose final output are generated requirements documents that integrate process model and execution-related data in an understandable fashion [20]. This approach consists of three main steps: 1) to analyze the process models that are relevant for the system, 2) to capture execution-related data, 3) to generate automatically requirements documents from the created models via a template-based natural language generation algorithm. Cognitive Theory of Multimedia Learning suggests that both textual and visual representations have to be presented at the same time to obtain high-quality requirements documents from process models. This approach includes 2 phases: preparation phase (identification of automatable activities, requirements analysis) and generation phase (sentence generation and refinement, document organization) [20]. E-company, e-trademark and e-government are three most popular project activities, which need generated requirements. However, this approach works only for process model, not datasets.

The electronic dissemination of financial and business information has developed significantly; the online reporting has evolved from a static PDF or HTML format to a dynamic XBRL format, which is expressed in 3 forms: concepts, relations and resources. Extensible Business Reporting Language (XBRL) facilitates the exchange of information and the standardization of financial information across multiple platforms and systems. The use of XBRL standard for business reporting becomes more feasible from various points of view: communication, data collection, control and analysis of financial information. An XBRL instance document is a business report published in a special format, which includes reporting period; amount of property, plant and equipment; amount of cash and cash equivalents; the amount of current inventories; annual income etc. [21]. Another important concept of XBRL as machine-readable 'dictionaries' is taxonomy as a 'dictionary of concepts and terms that a company needs to report its business information according to a set of standards or rules, which defines precisely these terms' [21]. Limitation of the study is that includes only the views of accountants and do not take into account the opinions of other stakeholders.

Firms explore the power of natural language processing (NLP) to automate processes and make data-driven decisions. Behera et al. propose a framework with the formulation of ethical principles to ensure NLP is safe, secure and reliable for responsible decision making and results in social benefits [22]. NLP is the process of understanding how computerized systems use texts, speech, and other resources, and how they are operated on computers. The major purpose of NLP is to achieve human-like language processing for a variety of jobs or applications, as well as to generate different reports using computational approaches to examine the generated texts. Business managers can apply NLP for a variety of organizational analysis, communications processing and textual analytics tasks. NLP, like any other technology, may be harmful to cause compromising privacy. Responsible decision-making is the ability of NLP systems to make constructive choices about business behavior based on ethical standards.

Because the relationship between products, companies and customers in the business environment become more complex and the life cycle of products or services shortens, traditional expert-centric methods become time-consuming and labor-intensive. As a result, managers and industrial practitioners are increasingly bracing a data-driven approach to access reliable results concerning business areas with opportunities. The implications of data-driven business opportunity analysis differ depending on data sources (e.g., both standardized and not standardized data) [23]. Language model is adopted with Choia et al. to effectively analyze numerous and variously written business items, and the local outlier factor method is utilized to measure novelty in a quantitative manner using different patterns [23].

Analyzing process data enables organizations to gain valuable insights into how processes are executed and what are possible improvement opportunities; process data querying allows analysts to explore the data with the intent of getting insights about the execution of business processes [24]. The current generation of process query languages focuses data scientists, however, there is a need to a query language to support domain analysts who may be inexperienced with database technologies, where the interface takes a natural language query from the user, automatically constructs a corresponding structured query to be executed over the stored event data [24]. Domain analysts and even end users should be empowered to benefit from the opportunities of process analytics in performing their activities in digital processes, because main limitation of process analytics technologies is that they do not make available process data accessible to human users in natural manner. Kobeissi et al. propose advantages of machine learning and rule-based approaches constructing content, behavioral, and performance queries [24].

The results of these materials served as a starting point for our research. Our developed product is unique because it can be used as a library and conveniently integrated into any existing system with Java support. For example, the company's website can be used not only for reports creating, but also for the convenient calculation of various metrics with subsequent saving in the database.

3 Methodology (Requirements)

As a foundution for this section we have analyzed diffrent business reports from various domains (e.g. finance: risk reports on loan obligations, travel: booking statistics in different periods, sales and promotions: website visit analysis based on promotional campaigns, etc.), research materials on this topic, and our 15 years of experience in IT across diverse business domains. We can apply data-driven approach and decision tree tool as methodology to develop business reports. Data-driven approach is an approach to business management and building analytical reports based on the use of big data. A decision tree is a decision support tool used in data analysis and statistics.

Lets start from comparing different business domains from data perspective. What is the difference?

1. Model. Each business works with its own unique data. For example, Healthcare: Patient information, including medical history, diagnoses, lab results, and treatment plans, as well as operational data on staffing, inventory, and billing. Manufacturing: Data on production processes, inventory levels, supply chain management, and quality control, as well as data on sales and customer demand. These data have its own unique types and structure. All of this is the model.
2. Storage and transmission. Data can be stored in databases, files, clouds, etc. Different protocols can be used to transmit them, such as FTP, HTTP, SOAP, and so on.

What unites different business domains?

1. Structuredness. Although the data themselves differ, they all have a certain structure and certain dependencies.
2. Transform data for various types of calculations as the most obvious thing.
3. Mathematical formulas are the form of presentation of absolute majority of calculations.
4. Reports are created on a regular basis.

Based on this, we can formulate the following list of requirements.

1. Support for obtaining a model from various sources: databases, files, online messages, etc.
2. Ability to process data. Since models are structured, they can be represented using any structured data exchange format such as JSON, XML, YAML etc.
3. The ability to dynamically change the model by changing and adding new data, with support for complex mathematical formulas and functions for their extension.
4. Return the modified model to the user.
5. The process of creating a new model should be as automated as possible and not require regular human intervention.

In this paper, we will consider the most interesting and probably the most challenging requirement: the need to change the model and add new data. Report generation is usually based on basic templates created by humans, and the report creation processes themselves are automated. Therefore, all rules for changing the model should be documented in a template, which after applying to the model will result in a new modified model that will serve as the basis for the report. Since this template will be developed by humans, all rules for changing the model should be documented in the most convenient form for humans. To do this, a GPL (general-purpose language) needs to be developed. Given that this tool is intended to be used in completely different business areas, the developed language should have elements of DSL (Domain-specific language) to be flexible and adaptable to the specific needs of the business.

Let's formulate the requirements for the GPL in more detail:

1. Concise and understandable for humans.
2. Adaptive to business needs.
3. Describe transformations.
4. Transformations should support various types of data: numbers, text, dates, sets of data, etc.

Currently, there are a huge number of programs and tools available for building reports, from asset management [25] to tenant screening services [26]. Most of them are narrowly focused and used in specific domains. However where are some exceptions like Microsoft Excel, Tableau, SAP Crystal Reports, JasperReports, QlikView. The question arises as to how well they meet the criteria we have identified for building reports and how they describe transformations. From the perspective of data transfer and acquisition, these tools are quite flexible, but in terms of supporting different data models, they are only partially effective. In most cases, these programs work with a fixed set of tabular data and sometimes require human intervention for their configuration and report generation, which adds significant complexity to the process automation. Additionally, tabular data is not flexible and has certain limitations. Data transformations are mostly done using mathematical functions and formulas. The variety of these rules is quite wide, but adding new transformation functions is a complex and non-trivial process and in most of the cases require programming skills [27, 28], and not all tools support these capabilities.

Let's consider the use of mathematical formulas of different complexity as a basis for a GPL from the perspective of the requirements we have formulated.

1. Conciseness and comprehensibility for humans. This point depends heavily on the complexity of the formulas and the knowledge of the individual. For example, simple calculation formulas such as Gross margin (Gross profit divided by revenue), Current ratio (Current assets divided by current liabilities), Net profit margin (Net income divided by revenue) are fairly straightforward. However, formulas for regression analysis and calculations of mature value (the value of an investment at a future point in time, based on a specified interest rate and time period) can be quite difficult to understand. However, with the support of creating custom functions, all complex calculations can be wrapped in a single function call with a set of parameters.
2. Adaptability for business needs. In this point, the need for DSL elements is revealed. Let's highlight two criteria that need to be considered:

- The first is the integration of the model into the mathematical formulas, namely the use of object names as part of the calculations. For example, instead of passing the actual values of Gross profit and revenue, we can pass their names as variables. Then the formula will look like this:

$$grossMargin = grossProfit/revenue.$$

This is more understandable and concise for humans. Since the input model usually has a complex hierarchical structure, symbols such as '.' or '−>' can be used to obtain sub-elements.
- The second criterion is the support for creating custom functions. Since particularly complex calculations can turn into huge formulas that are especially difficult to maintain and update. It is convenient to hide them behind custom functions that will only take the necessary arguments. Especially when such calculations are repeated in many places.

3. Transformation description can be presented as: $<place \quad in \quad the \quad model>=<transformation formula>$ To denote the place where the data should be placed after processing, the same approach with the use of variable names can be used. If priority is added to the calculations, the obtained results can be used in other calculations.
4. Support for different data types. Mathematical formulas work well with numerical data types but what about other types: string, dates, money so on. Such support should be implemented at the semantic level of the constructed GPL.

4   Results

Based on the above considerations, we now need to create a GPL utilizing mathematical formulas as the foundation. As an example, let us take basic calculations for an investment portfolio based on credit deals. We will examine different variants of the syntax, potential issues, and possible ways to address them.

The JSON data model will look like Listing 1:

```
{"Deals":[
    {
    "DealID":"deal1",
    "Loans":[
        {
        "Price":65,
        "LoanId":"1x1",
        "Balance":10000
        },
        .... ]
    },
    .... ]
}
```

Listing 1. Input model for credit deals for an investment portfolio

Each deal contains a set of loans with a price and balance. Task: build a report model with the following calculations:

1. Total balance for all loans
2. Total balance for each deal
3. Weighted average price by balance for each deal
4. Weighted average price by balance for each deal with a criterion price > 80
5. List of loan names separated by ‘,’

As a result, we should get the following model:

```
{
        "Totalbalance" : ***,
        "Deals":[
          {
          "DealID":"deal1",
                "Totalbalance" : ***,
          "WAP": ***,
          "WAP80": ***,
          "LoanNames":"lx1,..."
                        "Loans":[
                            {
                            "Price":65,
                            "LoanId":"lx1",
                            "Balance":10000
                            }  .... ]
          }, ... ]
}
```

Listing 2. Report model with the defined calculations

Let's see how our GPL will look using mathematical formulas:

1. **Total balance for all loans**. Probably the simplest calculation is just the total sum of objects in the set. Let there be a function SUM implemented in our system that takes a set of numbers as a parameter and returns the result in the form of numbers. The transformation rule will be as follows:

```
{
        "Totalbalance" : "SUM(Deals.Loans.Balance)"
}
```

2. **Total balance for each deal**. The calculation, although similar to the previous one, is much more complex for three reasons:

- The place in the model is not the root. As we discussed earlier, for a complex hierarchy, "." or "> "can be used.
- The result of execution will not be a number but a set of numbers. Let's simplify the task and assume that the SUM function takes a set of sets and returns the sum for each one.

- From the point of view of the GPL semantics, it is necessary to unambiguously identify that we need a subset of balances, not the entire series. Also, this problem needs to be viewed from the angle of possible business needs, namely the appending of additional filters and criteria. For example, for portfolio section, it is necessary to calculate the sum of balances only for loans with a higher average price or for loans with the price less than certain value.

Such requirements can be implemented in several ways:

- You can change the logic of *Deals.Loans.Balance* and instead of a single list, receive a list of lists. In this case, we need to add a function to combine lists (for example, to calculate *Totalbalance* for all deals). In addition, it should be remembered that the level of nesting can be much higher, and then we will already get a list of lists of lists. In this case, the merge function must have additional parameters describing the nesting level.
- In many structured languages, "[]" is often used to filter elements. In this case, to indicate that exactly at this level, we need to take a list of elements, we can make a mark "*[i]*" and then the formula will look like:

```
{
    " Deals.Totalbalance " : "SUM(Deals[i].Loans.Balance)"
}
```

Also, "[]" can be used to specify conditions for forming subsets. For example, *Deals.Loans[price > 80]*. Balance will return us a list of loan balances with a price greater than 80.

- The option can also be implemented at the functional level by writing a *getElements* function that will return lists of the necessary elements. In this case, the transformation formula will be…

```
{
" Deals.Totalbalance " :
"SUM(getElements(Deals.Loans.Balance)"
}
```

When necessary, an additional parameter can be added to the condition for forming a subset filter.

```
{
    " Deals.Totalbalance " :
     "SUM(getElements(Deals.Loans.Balance,price>80)"
}
```

3. *Weighted average price by balance for each deal*. The formula for finding the weighted average is the sum of all the variables multiplied by their weight, then divided by the sum of the weights. We already have the formula for calculating the

sum of the list, but to obtain the list we need to either multiply two lists, in which case we need to implement multiplication operation on sets, or get the list of multiplied elements. Since we are already calculating the sum of balances, we can use priority if it exists in our rules and use:

```
{
"Deals.WAP" :
"SUM(getElements(Deals.Loans,Balance*price)
/getElements(Deals, Totalbalance)",
"priority":2
}
```

We can also write a function to calculate the weighted average that takes variables and weights as parameters:

```
{
"Deals.WAP":
"WtAvg(Deals[i].Loans.Price,Deals[i].Loans.Balance)"
}
```

4. **Weighted average price by balance for each deal with the criterion price > 80** requires the implementation of conditions. As a result, we can obtain the following variants:

```
{
  "Deals.WAP80" :
  "SUM(getElements(Deals.Loans,Balance*price,price>80)
/SUM(getElements(Deals.Loans, Balance)"
}

{
 "Deals.WAP80":
"SUM(
Deals[i].Loans[price>80].Price*Deals[i].Loans[price>80].Balance)
/SUM(Deals[i].Loans.Balance)"
}

{
 "Deals.WAP80" :
 "WtAvg(if(Deals.Loans.Price>80,Deals.Loans.Price,0),
Deals.Loans.Balance)"
}
```

5. **The list of credit names separated by ','** is best done using the concatenation function:

```
{
    "Deals.WAP80"      :      "Concat(getElements(Deals.Loans,
LoanId),',')"
    }
```

Having a basic understanding of GPL raises the question of how to represent it in a programmatic form for convenient computation. Since mathematical formulas have different operation priorities (multiplication and division are performed before addition, parameters need to be computed before functions, operation priorities can be set using parentheses, etc.), there are many data structures that can be used for such computations. Since calculations take place from the highest priority operations to the lowest, a stack (a list that only allows access to the last added element) can be used, where operations with the highest priority will be at the end. Alternatively, trees (a data structure consisting of nodes that have references to child elements) can be used, where calculations will take place from bottom to top until they reach the root.

Let's consider what form our constructed GPL will have on such a data structure as a tree. To do this, we will describe how transformation rules will be transformed into nodes:

- Variables (<u>Deals.Loans.Balance</u>, <u>Deals.Loans.Price</u>, etc.), numeric constants (80,0), and text constants: since they do not require additional computations, they can be represented as nodes without child elements.
- Binary operations: addition, multiplication, division, subtraction, and comparison operations should be represented as nodes with two child elements.
- Functions will be transformed into nodes with a number of child elements equal to the number of parameters.
- Unary minus can be considered as a function with one parameter.

Let's look at some examples (Fig. 3, 4, 5 and 6): For the sake of clarity, in formulas we will use custom variable (CV).

Parsing formulas and converting them into tree structures constitute a fundamental component in the development of a tool that leverages the constructed GSL library to generate reports. This endeavor commences with the definition of the task at hand, followed by the identification of input and output parameters. Furthermore, an exploration of prospective alternatives and algorithms is undertaken to facilitate the achievement of our objective.

The input parameters encompass textual representations of formulas. These formulas may consist of the following components:

- Numeric values: 1, 3.14…
- Boolean values: true, false
- Constants: Pi, E…
- Variables and elements of the input data model: TotalBalance, Deal.loans.balance
- Text constants: 'Const text'
- Arithmetic operations with different priorities (multiplication and division have higher priority): $+, -, /, *$
- Parentheses for changing the priority of operations
- Boolean operations: &&, ||,…
- Comparison operations: ==, !=, >, >=…
- Standard functions: SUM, MIN, SIN, as well as custom functions for specific business needs with arbitrary parameters

The desired output is a data structure that allows us to:

Fig. 3. Simple arithmetic calculation: 2 + CV.

Fig. 4. Arithmetic calculation with default priorities: *2 + CV*1.1.*

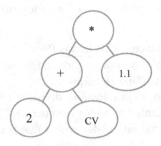

Fig. 5. Arithmetic calculation with changed priorities: *(2 + CV)*1.1.*

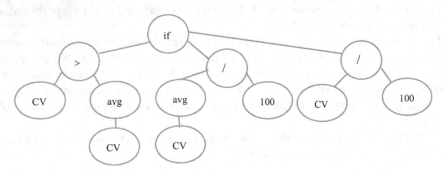

Fig. 6. Arithmetic with multiple functions: *If(CV > avg(CV),avg(CV)/100,CV/100).*

- Perform the specified calculations based on the input model and previous computations.
- Update the model with the computation results.

To solve this task, we can utilize libraries and systems for grammar creation and processing:

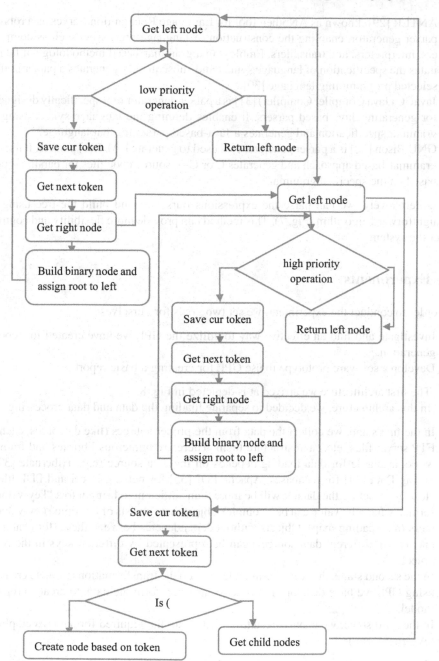

Fig. 7. Formula parser algorithm.

- ANTLR [29], known as ANother Tool for Language Recognition, serves as a robust parser generator, enabling the construction of language processors such as compilers, interpreters, and transpilers. Employing a grammar-based methodology, it facilitates the specification of language syntax and subsequently generates a parser in the selected programming language [30].
- JavaCC (Java Compiler Compiler) [31] is a parser generator tool specifically designed for generating Java-based parsers. It enables defining the language syntax using a grammar specification and generates a Java-based parser for that language.
- GNU Bison [32] is a parser generator tool used to generate LALR(1) parsers. It uses a grammar-based approach and generates C or C++ source code files for parsing input based on the specified grammar.

Alternatively, we can parse the expressions ourselves and build the tree using a straightforward algorithm (Fig. 7). This method can provide more flexibility and control over the system.

5 Experiments

In order to conduct the experiment, we set two goals for ourselves:

- Investigate and find an effective way to utilize the GPL we have created for report generation.
- Develop a software prototype to use GPL for creating a basic report.

The first architecture we arrived at is depicted in Fig. 8.
In this architecture, we decided to separate loading the data and data processing.

- In the first stage, we collect the data from the proper sources (like databases, cache, FTP server files, etc.) and structure them. There are numerous libraries and frameworks available for data loading depends on the data source (e.g., Hibernate [33], Spring Data [34] for databases, Apache POI [35] for parsing Excel and PDF files, etc.). To structure the data, it will be more than sufficient to bring it to a "key-value" format, where the value can be a complex object that consists of multiple "key-value" pairs (a cascading map). Objects without dependencies between them (for example taken from different data sources) can be represented by different keys in the root object.
- In the second stage, the software module utilized the transformation template created using GPL we have developed to process and transform the data, to create a report model.
- In the third stage, we convert resulting model into the required format (Excel, plain text, email, etc.).

The main advantage of this architecture is a security. As data collection and processing done on the server, we prevent user from accessing data beyond the report. Moreover, if the template for some reasons falls into unauthorized hands, it would not provide the intruder the information that could be used to obtain sensitive data.

However, this architecture has significant drawbacks:

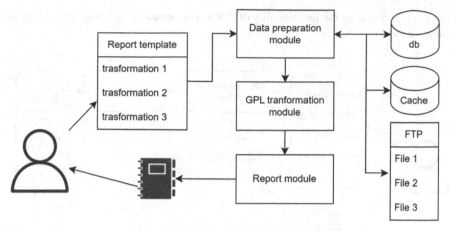

Fig. 8. Structure of program module

- Low flexibility. For each template, we need to maintain a list of required data sources, and if the template underwent significant changes, the need to modify the data sources accordingly.
- The complexity of creating reports significantly increases when there is a need to load additional information after certain transformations. For example, we have calculated general statistics for loan prices for a specific portfolio. If static shows significant profit drop we want to load a list of additional loans that could significantly improve the results and include it in the report.

As a result, we have come up with the second version of the architecture depicted in Fig. 9.

In this version, we have decided to combine the data load and data processing, allowing the user to specify the data sources, order of retrieval, and data processing in the template. In order for this architecture to be functional and efficient, we had to address certain challenges:

- Security: The template could fall into the wrong hands, so sensitive data cannot be stored within it.
- Standardization: The template had to be written in the GPL we have created and comply with its semantic and lexical standards, including the use of mathematical formulas as a foundation.

To address these issues, we have decided to use dynamic functions. As they are part of the GPL, they adhered to its standards. They are also sufficiently secure, as their implementation located on the server, and regular users have access only to the function name and passed parameters. With the correct level of abstraction, if all sensitive data are used exclusively within the function code and generalized information are passed through parameters, such an approach can be safely used for loading data.

Thus, we have completed the first part of the experiment, namely finding an effective way to utilize the GPL we have created for report generation. The second part involves demonstrating the functionality of this implementation, specifically by creating a software module.

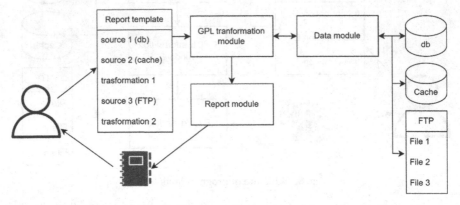

Fig. 9. Structure of program module

As input program module accepts report builder template, created using defined GPL and presented in JSON format. Template includes data source and list of data transformations (Listing 3), order of the commands will be considered as priority.

```
{
        "Loans" : "GenerateTestData('Loans','1x1','1x2','1x3','1x4')",
        "TotalBalance" : "SUM(Loans.Balance)",
        "WAP" : "SUM(Loans.Balance*Loans.Price)/TotalBalance"
}
```

Listing 3. Report builder template.

For presentational reasons we have used small JSON as a test data, however created tool can process big amounts of data. To show process more clearly we have transformed report builder template into list of commands (Listing 4).

```
@Test
void testReport() throws IOException {

    List<FormulaColumn> formulas = new ArrayList<>();
    formulas.add(new FormulaColumn( field: "Loans", formula: "GenerateTestData('Loans','1x1','1x2','1x3','1x4')"));
    formulas.add(new FormulaColumn( field: "TotalBalance", formula: "SUM(Loans.Balance)"));
    formulas.add(new FormulaColumn( field: "WAP", formula: "SUM(Loans.Balance*Loans.Price)/TotalBalance"));
    FormulaProcessor processor = new FormulaProcessor();
    print(processor.update(formulas).toString());

}
```

Listing 4. Report builder template presentetes as list of commands

Let us review the template presented on Listing 3–4. It consists of three commands:

1. Generating test data of loans with ids: lx1, lx2, lx3, lx4.
2. Calculating total balance of generated loans: *SUM(Loans.Balance)*.
3. Calculating weighted average of price by balance:

   SUM(Loans.Price* Loans.Balance)/ TotalBalace,

   where *TotalBalace* is a value calculated on the previous step. All commands from
template will be validated and processed and changed to trees (Listing 5–7).

```
Result:
▼ oo result = {FuncNode@6431} "FuncNode(funcName=GenerateTestData, params=[ConstNode(constName='Lo...  View
  ▶ 🌀 funcName = "GenerateTestData"
  ▼ 🌀 params = {LinkedList@6593}  size = 5
    ▶ ☰ 0 = {ConstNode@6595} "ConstNode(constName='Loans')"
    ▶ ☰ 1 = {ConstNode@6596} "ConstNode(constName='lx1')"
    ▶ ☰ 2 = {ConstNode@6597} "ConstNode(constName='lx2')"
    ▶ ☰ 3 = {ConstNode@6598} "ConstNode(constName='lx3')"
    ▶ ☰ 4 = {ConstNode@6599} "ConstNode(constName='lx4')"
```

Listing 5. GenerateTestData('Loans','lx1','lx2','lx3','lx4') presented as a tree.

```
Result:
▼ oo result = {FuncNode@6468} "FuncNode(funcName=SUM, params=[ConstNode(constName=Loans.Balance)])"
  ▶ 🌀 funcName = "SUM"
  ▼ 🌀 params = {LinkedList@6593}  size = 1
    ▼ ☰ 0 = {ConstNode@6598} "ConstNode(constName=Loans.Balance)"
      ▶ 🌀 constName = "Loans.Balance"
```

Listing 6. SUM(Loan.Balance) transformation presented as a tree.

```
Result:
▼ oo result = {BNode@6623} "BNode(operation=/, left=FuncNode(funcName=SUM, params=[BNode(operation=*...  View
  ▶ 🌀 operation = "/"
  ▼ 🌀 left = {FuncNode@6657} "FuncNode(funcName=SUM, params=[BNode(operation=*, left=ConstNode(con...  View
    ▶ 🌀 funcName = "SUM"
    ▼ 🌀 params = {LinkedList@6665}  size = 1
      ▼ ☰ 0 = {BNode@6667} "BNode(operation=*, left=ConstNode(constName=Loans.Balance), right=ConstNode(co
        ▶ 🌀 operation = "*"
        ▼ 🌀 left = {ConstNode@6670} "ConstNode(constName=Loans.Balance)"
          ▶ 🌀 constName = "Loans.Balance"
        ▼ 🌀 right = {ConstNode@6671} "ConstNode(constName=Loans.Price)"
          ▶ 🌀 constName = "Loans.Price"
  ▼ 🌀 right = {ConstNode@6658} "ConstNode(constName=TotalBalance)"
    ▶ 🌀 constName = "TotalBalance"
```

Listing 7. SUM(Loan.Price*Loan.Balance)/TotalBalance transformation presented as a tree.

After validation we will start executing commands line by line. The result of the first
command line would be a list of loans presented in a "key-value" format (Listing 8).

Listing 8. Generated test data.

```
@SpringBootTest
class ReportBuilderApplicationTests {

    @Test
    void testReport() throws IOException {

        List<FormulaColumn> formulas = new ArrayList<>();
        formulas.add(new FormulaColumn( field: "Loans", formula: "GenerateTestData('Loans','lx1','lx2','lx3','lx4')"));
        formulas.add(new FormulaColumn( field: "TotalBalance", formula: "SUM(Loans.Balance)"));
        formulas.add(new FormulaColumn( field: "WAP", formula: "SUM(Loans.Balance*Loans.Price)/TotalBalance"));
        FormulaProcessor processor = new FormulaProcessor();
        print(processor.update(formulas).toString());

    }

    ReportBuilderApplicationTests > testReport()
}

Report

✓ Tests passed: 1 of 1 test - 450 ms

Output:{
  "Loans": [
    {
      "Price": 65.0,
      "LoanId": "lx1",
      "Balance": 10000.0
    },
    {
      "Price": 65.0,
      "LoanId": "lx2",
      "Balance": 15000.0
    },
    {
      "Price": 80.0,
      "LoanId": "lx3",
      "Balance": 20000.0
    },
    {
      "Price": 39.0,
      "LoanId": "lx4",
      "Balance": 30000.0
    }
  ],
  "TotalBalance": 75000.0,
  "WAP": 58.6
}
```

Listing 9. Result of program module execution.

Executing second and third command lines will get result of requested calculations. After all command lines have been executed output JSON would be generated (Listing 9). It can be used to generate email report (Listing 10).

Credit portfolio

Number of loans:	4	
TotalBalance	$ 75,000.00	
WAP	$ 58.60	

Loan	Price	Balance
lx1	$ 65.00	$ 10,000.00
lx2	$ 65.00	$ 15,000.00
lx3	$ 80.00	$ 20,000.00
lx4	$ 39.00	$ 30,000.00

Listing 10. Report example (loans 1, 2 and 3 have value less than average WAP, loan 4 - more than average).

The result of the experiment shows effective way to utilize the GPL we have created to generate custom reports. We have created program prototype to show one of the examples of such implementation.

6 Conclusions

In this paper, we have formulated the foundation of GPL for creating dynamic reports. Additionally, we have examined mechanisms for processing such GPL for further use in calculating of various business metrics.

The results of the experiment show that by utilizing the approach we have described in the paper, it is possible to create a software module that can easily integrate with any business system. This module can be used for creating business reports and the ability to quickly change the template makes this process dynamic. The use cases of this approach are not limited to report generation alone. Complex mathematical calculations can be applied in various domains, (e.g. dynamically calculating the resources to reduce system workload, or dynamic calculations to determine pricing and discounts for business products and services etc.).

The construction of GPL for creating reports in various business areas is a very interesting and promising direction. Using mathematical approaches can make the process more flexible and adaptable to business needs. The huge advantage of such an approach is the conciseness and clarity of such a GPL for a human and the ease of its use. Among the disadvantages, one can highlight the difficulty of implementing such a language due to the numerous criteria that need to be considered.

References

1. Palmer, B.: What Are International Financial Reporting Standards (IFRS)? (2022). https://www.investopedia.com/terms/i/ifrs.asp/
2. Kobets, V., Yatsenko, V., Buiak, L.: Bridging business analysts competence gaps: labor market needs versus education standards. Commun. Comput. Inf. Sci. **1308**, 22–45 (2021). https://doi.org/10.1007/978-3-030-77592-6_2
3. Kobets, V., Yatsenko, V., Mazur, A., Zubrii, M.: Data analysis of personalized investment decision making using robo-advisers. Sci. Innov. **16**(2), 80–93 (2020). https://doi.org/10.15407/SCINE16.02.080

4. Savchenko, S., Kobets, V.: Development of robo-advisor system for personalized investment and insurance portfolio generation. Commun. Comput. Inf. Sci. **1635**, 213–228 (2022). https://doi.org/10.1007/978-3-031-14841-5_14

5. Kobets, V., Petrov, O., Koval, S.: Sustainable robo-advisor bot and investment advice-taking behavior. Lect. Notes Bus. Inf. Process. **465**, 15–35 (2022). https://doi.org/10.1007/978-3-031-23012-7_2

6. Kobets, V., Tsiuriuta, N., Lytvynenko, V., Novikov, M., Chizhik, S., et al.: Recruitment web-service management system using competence-based approach for manufacturing enterprises. In: Ivanov, V., et al. (ed.) DSMIE 2019. LNME, pp. 138–148. Springer, Cham (2019). https://doi.org/10.1007/978-3-030-22365-6_14

7. Kenton, W.: Business Segment Reporting Definition, Importance, Example (2021). https://www.investopedia.com/terms/b/business-segment-reporting.asp.

8. El-khoury, J., Berezovskyi, A., Nyberg, M.: An industrial evaluation of data access techniques for the interoperability of engineering software tools. J. Ind. Inf. Integr. **15**, 58–68 (2019). https://doi.org/10.1016/j.jii.2019.04.004

9. Lu, X.: Automatic analysis of syntactic complexity in second language writing. Int. J. Corpus Linguist. **15**(4), 474–496 (2010). https://doi.org/10.1075/ijcl.15.4.02lu

10. Hayes, A.: eXtensible Business Reporting Language (XBRL): Investor's Guide (2022). https://www.investopedia.com/terms/x/xbrl.asp.

11. Bondar, S., Ruppert, C., Stjepandić, J.: Ensuring data quality beyond change management in virtual enterprise. Int. J. Agile Syst. Manag. **7**(3–4), 304–323 (2014). https://doi.org/10.1504/IJASM.2014.065346

12. Nguyen, M.-T., Le, D.T., Le, L.: Transformers-based information extraction with limited data for domain-specific business documents. Eng. Appl. Artif. Intell. **97**, 104100 (2021)

13. Seng, J.-L., Lai, J.T.: An Intelligent information segmentation approach to extract financial data for business valuation. Expert Syst. Appl. **37**, 6515–6530 (2010). https://doi.org/10.1016/j.eswa.2010.02.134

14. Duque, J., Godinhob, A., Vasconceloscd, J.: Knowledge data extraction for business intelligence. Procedia Comput. Sci. **204**, 131–139 (2022)

15. Giner-Miguelez, J., Gómez, A., Cabot, J.: A domain-specific language for describing machine learning datasets. J. Comput. Lang. **76**, 101209 (2023)

16. Quintero, A.M.R., Pérez, S.M., Varela-Vaca, A.J., López, M.T.G., Cabot, J.: A domain-specific language for the specification of UCON policies. J. Inf. Secur. Appl. **64**, 103006 (2022)

17. Vidal, M., Massoni, T., Ramalho, F.: A domain-specific language for verifying software requirement constraints. Sci. Comput. Program. **197**, 102509 (2020)

18. Chavarriaga, E., Jurado, F., Rodríguez, F.D.: An approach to build JSON-based domain specific languages solutions for web applications. J. Comput. Lang. **75**, 101203 (2023)

19. Rodrígueza, A., Macíasd, F., Duránc, F., Rutle, A., Wolter, U.: Composition of multilevel domain-specific modelling languages. J. Logical Algebr. Methods Program. **130**, 100831 (2023)

20. Aysolmaz, B., Leopold, H., Reijers, H.A., Demirörs, O.: A semi-automated approach for generating natural language requirements documents based on business process models. Inf. Softw. Technol. **93**, 14–29 (2018). https://doi.org/10.1016/j.infsof.2017.08.009

21. Enia, L.C.: Empirical research: exploring extensible business reporting language and views of Romanian accountants. Procedia Econ. Finan. **32**, 1675–1699 (2015). https://doi.org/10.1016/S2212-5671(15)01495-1

22. Behera, R.K., Bala, P.K., Rana, N.P., Irani, Z.: Responsible natural language processing: a principlist framework for social benefits. Technol. Forecast. Soc. Chang. **188**, 122306 (2023). https://doi.org/10.1016/j.techfore.2022.122306

23. Choia, J., Jeong, B., Yoonc, J.: Identification of emerging business areas for business opportunity analysis: an approach based on language model and local outlier factor. Comput. Ind. **140**, 103677 (2022). https://doi.org/10.1016/j.compind.2022.103677

24. Kobeissi, M., Assy, N., Gaaloul, W., Defude, B., Benatallah, B., Haidar, B.: Natural language querying of process execution data. Inf. Syst. **116**, 102227 (2023). https://doi.org/10.1016/j.is.2023.102227

25. Best, R.: Best Asset Management Software (2023). https://www.investopedia.com/best-asset-management-software-5090064

26. Carmody, B.: Best Tenant Screening Services (2023). https://www.investopedia.com/best-tenant-screening-services-5070361

27. Kenton, W.: Visual Basic for Applications (VBA): Definition, Uses, Examples (2022). https://www.investopedia.com/terms/v/visual-basic-for-applications-vba.asp.

28. Hicks, M., Levin, D.: CMSC 330: Organization of Programming Languages (2013). https://www.coursehero.com/file/178765173/org-of-Progpdf/

29. ANother Tool for Language Recognition. https://www.antlr.org/documentation.html. Accessed 29 May 2023

30. ANTLR. https://github.com/antlr/antlr4. Accessed 29 May 2023

31. JAVACC, https://javacc.github.io/javacc/documentation/. Accessed 29 May 2023

32. GNU Bison. https://www.gnu.org/software/bison/. Accessed 29 May 2023

33. Hibernate. https://hibernate.org/. Accessed 29 May 2023

34. Spring Data. https://spring.io/projects/spring-data. Accessed 29 May 2023

35. Apache POI. https://poi.apache.org/. Accessed 29 May 2023

Small Businesses Participating in Digital Platform Ecosystems - A Descriptive Literature Review

Lukas R. G. Fitz[✉] [iD] and Jochen Scheeg

Department of Business and Management, Brandenburg University of Applied Sciences,
Magdeburger Str. 50, 14770 Brandenburg an der Havel, Germany
{fitz,scheeg}@th-brandenburg.de

Abstract. The emergence of digital multi-sided platforms has disrupted day-to-day business in many industries. Nowadays, many small businesses participate in digital platform (DP) ecosystems and act as value creators or complementors. Investigating the small business perspective does not have a rich tradition in platform research yet. In response, this paper presents a systematic literature review (SLR) analyzing the thematic points of focus and highlighting new avenues for information systems and small business research. We find that objectives related to digital transformation, digital strategy- and business model-development, and DP adoption are fundamental topics in the conjunction of small business and DP ecosystems. Based on the findings, an agenda for future research is developed, including propositions for practice-oriented and human-centered approaches fostering sustilience, innovation capabilities and digital leadership competencies in small businesses facing DP ecosystem participation.

Keywords: Digital Platforms · Business Ecosystems · Small Business · SME

1 Introduction

Ever since the great upscaling potential of multi-sided business models was leveraged by big tech companies for the creation of digital platform-driven ecosystems, researchers and practitioners in the fields of management, entrepreneurship and information systems (IS) have developed concepts and typologies to describe characteristics, mechanisms, strategies and actors in the digital platform (DP) economy. DP business is shaped by very high market dynamics and driven by impactful network effects [1–5]. Concerning strategy and business model development, incumbent views on industry rivalry, value chains and competition [6–8] have been complemented by modern business literature speaking of networked business ecosystems [9], platform revolution [4], innovation paths towards becoming ecosystem drivers [5] or blue ocean shifts [10].

Nevertheless, small brick-and-mortar firms with linear value chains, which are often paraphrased as "traditional business", constitute the foundation for many successful DPs, as they co-create value for the network [11, 12]. This is especially true for two-sided platforms, where small businesses belong to the goods- or service-providers' side,

J. Maślankowski et al. (Eds.): PLAIS EuroSymposium 2023, LNBIP 495, pp. 38–55, 2023.
https://doi.org/10.1007/978-3-031-43590-4_3

but also for larger multi-sided platform models in which they serve as complementors [13, 14]. The relevance of studying small businesses as actors in DP ecosystems can be derived from two general observations: (1) reaching a critical mass of complementors or users on all sides of a multi-sided DP is an important requirement for DP providers aiming to establish indirect network effects and to drive an ecosystem [15]. (2) Small businesses make up the largest proportion of enterprises across the globe, accounting for most employment and adding the most value [16]. Thus, small businesses engagement is an important success factor for many DP ecosystems, because they provide a large value potential for upscaling DPs to leverage.

This paper investigates small businesses' being and doing in and around DPs with an aim to depict the current state-of-the-art and elicit future research directions. The novelty of this research motivation is to center the study specifically around *small* providers and complementors within DP ecosystems. Correspondingly, Drechsler et al. [17] advocate for a new way of looking at small businesses in IS research, which is not only driven by size and turnover figures, but takes limited networking and innovation capabilities, limited resources and lower digital maturity levels into account. According to the authors, "it can be assumed that digitalisation in networked SMEs actually has different characteristics than in large enterprises and should be studied separately" (p.11). At the same time, we acknowledge that there exist SME definitions for companies with up to 500 employees [18]. After all, knowledge-based work is especially important for the emerging conjunct research area of small business [19] and IS [20].

2 Background

Using DPs and DP ecosystems is a technological mean to facilitate digital transformation (DT) initiatives [21, 22]. Such DT initiatives have lately been examined with a focus on ambidexterity in small and medium-sized enterprises (SME) [23–25]. Ambidexterity is the capability of a firm to exploit their business model towards higher efficiency and explore towards long-term innovation [26]. Results suggest that an exploration orientation positively moderates the relationship of platform/network capabilities and SME performance [24]. In addition, Heikkilä et al. [27] analyzed SMEs' digital innovation paths in relation to their orientation and identified three archetypes: (1) profitability seekers improving backend operations and use of resources first and find new channels last; (2) growth seekers improving customer targeting first and find new partners and channels last; (3) new business starters improving their business models as a whole and testing viability. Although all three paths provide opportunities for leveraging value from DP ecosystems, either at later or at earlier stages, it remains unclear under which considerations small businesses navigate any of these paths.

After all, regardless of ambidexterity capabilities, the heterogeneity of available DP solutions [28] makes it increasingly difficult for business owners to choose the right solution [29]. On top of that, previous small business research has shown how difficult it might be to understand small businesses' strategic decision-making: Small business leaders usually look out for DT solutions and seek for collaborations in the first place, which constitutes an orientation phase in DT initiatives [30]. This may even lead to innovation projects with external partners and research institutions, which was found to be beneficial [31, 32]. Nevertheless, human factors like individual motivation and emotional bias have a significant influence on decision-making [33, 34], since many small businesses are run by single or few business owners only, leading to an even greater weight of biased decisions [35, 36]. These characteristics underline the merit of studying small businesses separately [17], also in the context of DPs.

Analyzing the platform research landscape, De Reuver et al. [37] identified several concepts in IS literature, "Digital platform (sociotechnical view)" and "Ecosystem (organizational view)" being two of them. In particular, the authors criticize that these concepts lack uniform definitions around platforms, ecosystems and units for analysis. Furthermore, they object that digitality had played a minor role in the discourse on platform business models. In terms of scoping, De Reuver et al. recommended to widen the scope of DP research and develop more contextualized theories. In other words, a big picture of the so-called "platform revolution" [4] would be incomplete if individual perspectives, such as those of small businesses, were left out of scope.

An OECD report [29] resonates that a research gap exists for small businesses' ways of doing business via DPs, especially against the backdrop of a still-growing DP economy [1]. The authors criticize that little is known about internal views and themes associated with participating in such ecosystems as a small player. In consequence, we formulate our research question (RQ) as follows: *What characterizes small businesses' participation in DP ecosystems?*

3 Methodology

This study tackles the RQ by shedding light onto existing knowledge through a systematic literature review (SLR). It is thematically centered around the interplay between small businesses and DP ecosystems and aims to point at potential future research avenues that may enhance our understanding of small businesses' behavior and needs in this area. The motivation for conducting an SLR is twofold. First, SLRs play a crucial role for the further development of the small business research field [19]. Secondly, we envision this SLR to foster an emerging research direction with the aim to develop new theory, artifacts and practice-oriented contributions in the conjunct area of small business, IS and platform research. In particular, this article presents a descriptive review [38]. Such SLRs "employ structured search methods to form a representative sample of a larger group of published works that are related to a particular area of investigation" [38] (p.186). Altogether, the goal of this paper is to (1) provide an overview of existing works regarding small business and DP ecosystems with an emphasis on the most frequently discussed contents; (2) provide an analysis of the results and discuss emerging research gaps; (3) formulate propositions for new research directions from the current viewpoint. Our approach is documented in a rigorous manner to share our review experiences and increase transparency of our approach [39]. In addition, we follow academic guidelines for conducting SLRs in IS [40], Business [41], and Entrepreneurship [42] research.

Planning and designing the SLR, in particular explaining its relevance and scope, is a common starting point of all guidelines considered. At this point, we have already explained our scope and research motivation. We have also highlighted the value of this contribution against the backdrop of existing literature in the field. Therefore, we consider the first step to be completed. Next, we follow vom Brocke et al.'s [39] suggestion to further conceptualize our topic. Hence, we define search activities that will lead to actually conducting the search, selecting the sources, and identifying the relevant outcome in the subsequent section. We also describe sources taken into account as well as search strings and some descriptive statistics of the outcome. Implicit steps concern the paper extraction process, including selection criteria, analysis steps and final selection outcome. Finally, the last phase comprises writing down the results, discussion, and presenting an outline of future research opportunities.

The systematic search was conducted throughout a variety of journals and conference proceedings appearing in the "VHB-JOURQUAL-3" (JQ3) ratings with a rating of A +, A, B, or C to allow for a both broad and well themed, yet high-quality and impactful pool of sources [42]. To filter for most relevant sources in the light of our research scope, the JQ3 categories "Business Information Systems (WI)", "Technology, Innovation and Entrepreneurship (TIE)" and "Small and medium-sized enterprises (KMU)" were considered. Journals with B or C ratings were additionally filtered according to journal titles and descriptions pointing at the research fields of Small Business or Electronic Business or Business/Management Information Systems. Search procedure and outcome are summarized in Fig. 1. Because some of the selected journals are allocated in IS research and others in small business or entrepreneurship research, the search string had to be suitably formulated for both kinds of sources: *"(((((Micro OR Small) AND (Business OR Enterprise)) OR SME OR MSE OR MSME) AND (Platform OR Ecosystem OR Network)) OR ((Digital OR Online OR Virtual OR Electronic) AND (Platform OR Ecosystem OR Network))"*.

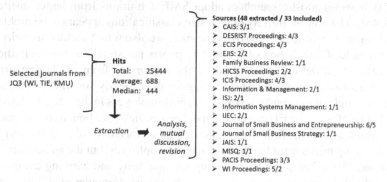

Fig. 1. Literature search procedure and outcome

Search operations were carried out in the selected journals' online repositories. After each search operation, all results were sorted by relevance and the top 50 results were scanned for potential extraction. The number of extracted articles each was documented in the end. The criteria for extraction were determined as follows: only academic and peer-reviewed research papers; only in English language; no publication older than 2010 – referring to the time when Yoo et al. [43] pointed at the peculiarities of innovation associated with DPs and the relevance to study them in IS research; title or abstract pointing at *both* small business and DP ecosystems. At this point, we consciously decided to exclude papers from general platform research, as our RQ deals with small businesses specifically and we intend to generate insights on this perspective. In the end, 48 papers had been extracted and were scanned thoroughly. Following this in-depth analysis, 15 more papers were excluded from the collection due to unfitting parameters regarding the before-described selection criteria. Details of remaining papers were collected in a spreadsheet. Therein, content summaries and classifications were added with a focus on four aspects: firm size considered in the paper, DP type considered, and key findings regarding small businesses and DP ecosystems. To limit individual bias, the literature search and extraction steps were conducted by two authors independently, who finally harmonized their results by consensus.

4 Results and Analysis

Following a concept-centric analysis approach [44], all key findings of the remaining 33 papers in the final selection were noted and clustered into qualitative thematic categories, allowing for more structured analysis and writing. The categories were subdivided into small-business-related categories and DP-related-categories. If applicable, one paper could be allocated to two categories at most. The synthesis outcome was summarized in a matrix, see Table 2. For better readability of the matrix, an ID is assigned to each paper in Table 1.

Firm Sizes. Apparently, there exists no consistent classification of "small business" in the research community, although the definition of "small" in most cases refers to the number of employees. In exceptional cases, regional authorities' definitions are adopted (e.g. [45]), however most researchers adopt SME definitions from larger institutions, (e.g. [18, 46]). The non-uniformity of firm size classifications appears to be problematic for producing transferable research outcome since we also found a wider disparity in our literature sample. Table 2 shows the count of papers per small business classification, derived either from definitions adopted by the authors or from the largest firm size considered in a case study. For instance, Spriggs et al. [47] study *small* family firms" and work with a sample of companies having max. 500 employees in the US, which would be already double of what the EU commission defines as the upper boundary for "medium-sized". In another case, the firm sizes focused-upon seemed to be clear at first, with 90% of the cases being micro firms having less than 10 employees, but the authors also stated that 10% had "10 +" employees, providing for insecurity and harming comparability of their study [48]. Being more transparent with outliers, Berendes et al. [49] include a company with 1000 employees and another one with 600 in their interviews, while 17 other participants had only up to 18 staff members. Although micro enterprises make up the largest proportion of SMEs in various statistics [50, 51], only two papers particularly addressed micro firms with less than 10 employees [52, 53]). Furthermore, studies that did not disclose the authors' small business definition at all were an even greater issue. We found that this applied to more than 50% of the papers in our literature collection. While these authors make use of common terms such as "SMEs", "micro", "small", they do not give away the exact firm size they are addressing. This is especially astonishing in regards of case or field studies where participants are portrayed as SME representatives but firm size details consequently remain obscure (e.g. [54, 55]).

Platform Types. Not many papers deal with DPs as a general phenomenon. The few examples found are commonly theorizing on the relation between small businesses' capabilities, strategic approaches or performances against the backdrop of static and dynamic DP characteristics [56–60]. Most authors however apply a narrower scope, considering specific kinds of DPs. By a large margin, e-commerce platforms for transactions involving goods and services are concerned in the literature collected, with a great emphasis on retail. In papers dealing with e-commerce as a research area itself, authors tend to speak of DPs as a "third-party" [53, 54, 61, 62]. Within this area, a special emphasis on local owner operated retail and local platforms [49, 54, 63] and combinations of e-commerce and social media platforms [64, 65] could be observed. However, social media platforms are also considered an own DP category. Despite the greater attention for B2C small business, DPs addressing B2B demands were also covered by a variety of papers. Two of these were among the design-oriented papers, developing DP concepts for value co-creation among small craft [66] and textile [67] businesses. Camposano et al. [68] provided insights on the perception of small actors in an emerging enterprise-to-enterprise platform for the construction industry. Overall, the connecting ecosystem function for value co-creation and networking [47, 55, 69–71] seems to play an overarching role around B2B-related DPs for small businesses.

Fields of Interest Related to the Small Business Perspective. The research motivation of most papers in the final basket is to study small businesses as dependent entities being impacted by challenges arising from the emergence of DPs. Therefore, the authors' fields of interest are focused on managerial aspects like strategic direction, objectives, behavior, decision-making or viewpoints existing in these companies. A major objective in this regard is the DT of small businesses, including digitalization and digitization measures. Being a well-discussed research field in itself, DT often goes along with new strategies featuring the innovation of business models, processes and methods, technology adoption and organizational capabilities, also connected to human skills and knowledge, among others. One response to the urgent need of many small businesses to digitally transform in such ways is, looking at 17 papers in the selected literature pointing that way, participating in DP business ecosystems. Some of these papers encompass DPs as a novel opportunity for incumbent small businesses to implement DT, hence as enablers for digital strategy and business development [48, 52, 54, 59, 63, 69]. Some studies take the strategic consequences of DPs for small businesses into account but omit the focus on transformatory effects, as they rather examine performance indicators [45, 74] or human factors in the entrepreneurs' perception of DPs [65, 70]. Other authors observed DT effects through DP referring to small businesses' web presence maturity [52, 57, 76] and capabilities relevant for innovation [62], such as managing and sharing knowledge [71]. Innovation capabilities also played a decisive role in nine papers of our collection, for instance connected to small business organization and leadership [47, 56, 75] or performance like marketing success [61] or innovation speed [60]. Few papers in our selection were written from a DP provider's perspective, with small businesses being either the target or source of impact or both. We found two streams in this direction: (1) papers analyzing existing [72, 73] or designing new DPs [67, 80] tailored specifically for small businesses; (2) papers evaluating the value created by small businesses when using DPs in the role of complementors [58, 64, 70, 72, 77].

Table 1. IDs assigned to papers selected for analysis

ID	Reference	ID	Reference	ID	Reference
1	Foster & Bentley, 2022 [72]	12	Holland & Gutiérrez-Leefmans, 2018 [73]	23	Marheine et al., 2021 [70]
2	Mkansi, 2021 [53]	13	Pfister & Lehmann, 2022 [74]	24	Hiller et al., 2020 [69]
3	Pan et al., 2022 [62]	14	Bollweg et al., 2020 [54]	25	Hönigsberg, 2020 [67]
4	Wirdiyanti et al., 2022 [45]	15	Omotosho, 2020 [65]	26	Ha et al., 2016 [64]
5	Mandviwalla & Flanagan, 2021 [48]	16	Karjaluoto & Huhtamäki, 2010 [52]	27	Asadullah et al., 2020a [75]
6	Burgess, 2016 [57]	17	Madill & Neilson, 2010 [76]	28	Bärsch et al., 2019 [63]
7	Spriggs et al., 2012 [47]	18	Santoso et al., 2020 [77]	29	Deilen & Wiesche, 2021 [58]
8	Benitez et al., 2022 [56]	19	Li et al., 2011 [78]	30	Heimburg et al., 2021 [55]
9	Wu et al., 2022 [60]	20	Camposano et al., 2021 [68]	31	Asadullah et al., 2020b [61]
10	Li et al., 2018 [79]	21	Bartelheimer et al., 2018 [80]	32	Gierlich-Joas et al., 2019 [59]
11	Scuotto et al., 2017 [71]	22	Berendes et al., 2020 [49]	33	Rauhut et al., 2021 [66]

Fields of Interest Related to the DP Ecosystem Perspective. Besides their focus on small businesses, the papers collected also deal with DP-related fields of interest that were categorized in this SLR. From an individual firm's perspective, joining a DP requires change. Technology-driven change is a popular topic, especially in IS literature, to be examined under the aspect of technology adoption. This concerns for example antecedents of DP adoption, such as decision factors [78] or adoption requirements [66], expectations within the adoption phase [49] as well as post-adoption impacts [45]. In some cases, DP adoption is not viewed as an isolated process but goes along with other DP activities of small businesses. For instance, some authors consider co-creation and co-operation within a DP ecosystem as a value-adding implication or a side aspect of DP adoption [69, 70, 74]. DP adoption was also studied in conjunction with communication capabilities of small businesses [57, 65], work practice and organizational learning [61, 75, 78], Covid-19 crisis response [48], or in consideration of DP ecosystem design requirements [49, 59, 68]. DP-induced value co-creation was also taken into account as a capability in itself [47, 60] or in combination with organizational learning and

Table 2. Literature analysis matrix with categorized papers; [1]Digital Strategy / Business Model Development, [2]Innovation Capabilities, [3]DP design for small business requirements

Category/Paper ID		1	2	3	4	5	6	7	8	9	10	11	12	13	14	15	16	17	18	19	20	21	22	23	24	25	26	27	28	29	30	31	32	33	Σ
Firm size	not disclosed	X		X	X	X				X	X		X		X	X		X	X		X	X	X				X		X	X	X			X	*18*
	≤ 10		X														X																		*2*
	≤ 50						X					X												X		X						X			*5*
	≤ 250																								X			X					X		*3*
	≤ 500 / ≤ 1000*					X		X	X					X						X															*5*
DP types	E-Commerce	X		X	X		X			X	X			X	X	X	X	X	X		X	X	X	X	X				X			X		X	*20*
	Industry / B2B		X			X		X				X														X				X	X		X		*8*
	All/General												X							X				X						X	X				*5*
	Social Media					X			X	X													*X				X								*4*
	Local													X	X														X						*3*
Content: DP ecosys	DP Adoption	X		X	X	X	X				X		X		X	X	X	X			X	X	X		X	X	X	X	X		X	X	X	X	*21*
	DP Design											X			X	X	X		X		X	X	X	X	X	X	X			X	X		X	X	*10*
	Co-Creation						X	X		X		X							X	X			X	X	X	X				X	X				*9*
	Work/Learning								X											X							X	X							*6*
	Communication						X																				X	X							*3*
	Crisis Response																												X						*1*
Content: small business	DT	X	X	X	X	X	X	X	X	X	X			X	X		X	X			X			X	X				X			X	X	X	*17*
	Dig.Strat/BMD1	X	X	X	X	X	X	X							X	X	X	X		X	X	X	X	X	X				X				X		*14*
	Innovation cap.2			X			X			X								X				X						X			X	X			*9*
	DP des. For SBR3	X						X					X											X	X	X					X				*7*
	Value Creation																		X								X			X			X		*4*
	Dig. Leadership						X	X	X							X								X				X							*4*
	Performance			X	X					X			X	X																	X	X			*4*

development [71]. DP use for value co-creation and networking also motivated design-oriented research and evoked conclusions pointing at design-relevant insights [55, 67, 77].

Synthesis of Cross-Dimensional Foci. A synthesis based on the co-occurrence of small business and DP-related content in the reviewed literature allows for an analysis of cross-dimensional foci. The analysis outcome is depicted in Fig. 2. Considering the extensive coverage of DT and DP adoption in the analyzed content, it is unsurprising that the overlap between both dimensions forms a major focus area (14 papers). One could argue that both activities mutually depend on each other, at least to a certain extent: DT may require adoption while adoption may imply DT. It is this very interplay, which apparently yields potential for researchers in the practical arena of small businesses acting in DP ecosystems. DP adoption is also entangled with the second biggest cross-dimensional focus (13 papers), digital strategies and business models in small businesses. Here, the interplay continues: DT, anyhow motivated, requires digital strategy, strategy needs to consider DP adoption requirements, while DP adoption also requires a strategy itself. At the same time, DP adoption is a DT initiative and DT eventually affects the business model. Eight papers were taking all three aspects into account and point at such interrelationships. For instance, Mandviwalla and Flanagan [48] see DP adoption as a mean to accelerate DT; they however state that it was challenging especially for micro businesses "to identify the relevant trees - solutions - in the forest of digital options" (p. 369). Karjaluoto and Huhtamäki [52] suggest that small businesses should not blindly aim at the highest degree of e-channel utilization, which is platformized e-commerce, but should rather evaluate the role of e-channels strategically, in consideration of internal and external influence factors. With a special focus on local e-marketplaces, Bärsch et al. [63] question sustainability of DP utilization. Nevertheless, they highlight opportunities to lower DP entrance barriers by beginning with and learning through *local* platform adoption. Still, the entrance barriers may be hard to overcome. Small business owners could therefore benefit from strategy development support, for instance sponsored by cities [54]. In sum, the significance of interrelations between DP-induced DT, DP-induced strategy/business model transformation and DP adoption is emphasized by the overlap of thematic dimensions in this analysis.

Besides, leveraging value from these interrelations cannot go without identifying resources and capabilities at hand [81]. Following the cross-dimensional content analysis, innovation capabilities are often linked to DP adoption, too. Moreover, they may be connected to work and organizational learning in small firms and also to co-creation and co-operation among DP complementors. Another focal point can be derived from two dimensions adopting a rather platform- and design-driven view: analysis/design of DP ecosystems for small businesses and viewing small businesses as value-creators in a DP-based value network. Apparently, both are connected to the objective of designing DPs, but digitality and small business orientation form a new context here. In response to the vast and overwhelming variety of solutions, it is not only the SME decision-makers' task to select the right DP solution, but also the DP providers' task to create meaningful and low-boundary offers. Taking findings on SMEs' contribution and roles in value networks [58] into account, small actors seem to provide a lot of potential for new design requirements when developing DPs in a long-term perspective. As Gierlich-Joas

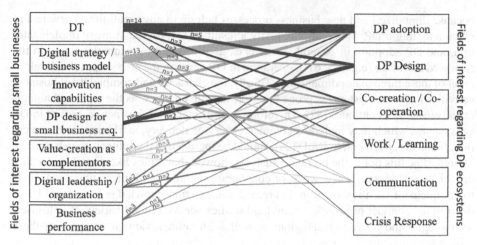

Fig. 2. Thematic co-occurrences in reviewed literature; n = amount of papers

et al. [59] put it: "Not only strategies for the firms' development need to be defined, but also strategies for the growth of the whole ecosystem are required" (p.12). The cases presented by Camposano et al. [68], both from the B2B construction sector, report that it took up to 5 years before small firms started leveraging real advantage from the co-creation opportunities in the examined DP ecosystem. Nevertheless, these firms were already able to find strategic partnerships and build trust in an early stage of an emerging DP. Overall, these insights stress the necessity for strategic and sophisticated DP design, also regarding the DP's ecosystem.

5 Development of a Research Agenda

Drawing upon the present literature analysis, three major interrelated focal points are highlighted: (1) DPs as an element of small business DT initiatives, (2) digital strategy and business model development in small businesses involving DPs, and (3) DP adoption by small businesses. The domination of practice-oriented papers in the research field, especially case studies, provides a valuable pool of exemplary cases to be exploited by both scholars and practitioners. On the other hand, it seems like the research field is lacking profound theory, especially regarding constraints in small businesses' strategy development and decision-making. Case after case, the studies' limitations reveal a lack of understanding for the actual "Why" regarding small businesses' participation in DP ecosystems. At the starting point of proposing directions for a future research agenda and as a finding towards the RQ, the three major focal points may be considered as fundamental and interrelated elements to be considered in platform research with a small business context. To encompass the full landscape of factors driving the participation of small businesses in DP ecosystems, surrounding topics should be included in the research agenda, which will be discussed in this section.

Strengthening the Evidence Base on Small Businesses' DP Use. To begin with, it is important to consider day-to-day DP-based actions when studying small actors in DP, because digitalized and new business processes belong to any small firm's new reality when adopting a DP. This reality is firstly a result of DT, digital strategy implementation, and DP adoption in general. Vice versa, as illustrated by examples found in the reviewed literature (e.g. [55, 61]), operating in DP ecosystems also shapes the development of these three major dimensions. Thirdly, the character and complexity of business activities in DP ecosystems is prone to unforeseeable changes, as the rapid technological advancement continues [82]. Therefore, operational small business activities in DP ecosystems, which may include e-commerce, service delivery, marketing, communication, financing, payment, or R&D [29], should be included in the future research agenda. To underline this point, the literature analysis showed that topics like digital leadership, organizational learning, new work, or collaboration could yield interesting dimensions to complement such research. In this regard, the existing knowledge base could benefit from the conduct of more case and field studies across a broad variety of industries, technologies and geographical regions, as well as literature reviews accompanying these efforts.

Competency Building and Human-Centeredness in DP Adoption Support. It is not centrally necessary for entrepreneurs to be experts in all disciplines if they find suitable ways to compensate lacking competencies [83]. However, awareness and education around DP ecosystems embedded in digital leadership may be decisive factors. To our best knowledge, the research community has not yet come up with competence-building measures tailored for small business executives wanting to implement digital leadership in DP ecosystems. In fact, when it comes to the rise of the DP economy, small businesses are likely to get stuck in a "strategic crisis" where urgent action is required but the organizational adaptability is low [84]. In theory, organizational readiness for change [85] is a very human-driven influence factor, shaped by human motivations, perceptions, behaviors, beliefs, et cetera. The reality in most small businesses' day-to-day operations and decision-making processes apparently reflects this theory well [86], which makes it necessary to consider these aspects when dealing with small firms in research [17] and practice. Because DT initiatives often feature maturity or stage models designed for larger companies, which was also criticized in here-reviewed literature regarding small firms' web presences [52, 57], problem inquiry phases often end without any human-centered investigation [87]. Taking specificities of small firms into account, we suggest to consider organizational readiness for change with human-centered import as another influence factor for small businesses in DP ecosystems. Except for recommendations on perspectives to be applied in field research [17], the prevalent literature base is missing propositions on how to accomplish readiness for change in small firms entering DP ecosystems.

Sustilient DP Design for Small Businesses. Especially in the design-oriented research direction, the current state-of-the-art indicated several challenges in developing new DPs that are trustworthy and easy to adopt for small complementors. We adopt the neologism "sustilience" (it combines sustainability and resilience), introduced by Grant and Wunder [84], and propose sustilient DP design as an emerging research agenda element demanding for further investigation. It might also be of scientific interest to further enhance the concept of sustilience in the general context of small businesses. This could help facilitate a future, in which UN sustainability development goals [88] play an important role. Furthermore, considering the increasing use of AI in DPs [89], trust, reliability and safety [90] may serve as paradigmatic values for small business-centered DPs, because many small firms are economically incapable of trying out unreliable or risky solutions. This calls for "sustilient" actions on both sides: a) small businesses need to steer their whole business transformation sustiliently, while DP providers need to b) design sustilient DPs and c) enable sustilient change for small firms adopting DP.

Innovation and Leadership Capabilities for DP Use in Small Businesses. In the context of DP ecosystems, innovation capabilities are innovative solution-finding skills, innovative structure, organization, work processes and culture, but also the management of resources like digital assets [91–93]. One problem is that small business owners are likely to have limited overview of such capabilities, since they are rather experts in their original professional domain [48]. Furthermore, networked innovation in an innovation ecosystem encompasses newly emerging methods and ways of thinking, such as "horizontal and inclusive economic thinking, as well as enabling certain organizational continua, relevant for interactive innovation and dispersed patterns of production" [94](p. 5251). Furthermore, our literature analysis shows that innovation capabilities co-occurs with aspects of digital leadership and organization or performance aspects. Heimburg et al. [55] observe that auxiliary services provided by DPs help professionalizing small firms in coping with the whole new ecosystem and benefitting from it. Such findings add another cornerstone to the bigger picture: Developing innovation capabilities and being able to *lead* a small business in DP ecosystems, in contrast to managing it [95], are competencies that play a primary role in surviving as a small platform actor. Consequently, as another element of the new research agenda, we suggest a promising research direction towards professionalizing small business actors towards the development and exploitation of innovation capabilities and resources, which are suitable for self-efficacious acting in DP ecosystems.

6   Limitations

Although the selection of papers found in journals and proceedings contributed in a valuable manner to our literature review, there are obviously limitations affecting our study. First of all, the research landscape around both small businesses and DP ecosystems provides a broad variety of terms and keywords, which are interchangeably used by the research community. Therefore, we must assume that our formulated search string could not reach all relevant hits possible in the sources considered. The JQ3-based variety of considered journals and proceedings helped to maintain a thematic focus, but also limited the reach of our search activities. On top of that, we decided against a more exhaustive search covering platform research articles without small business import, despite being aware of "broader" literature on DP complementors. Future literature reviews may choose another direction and derive insights from papers concerning bigger companies, too – however, the variety of study backgrounds and small business conceptualizations already inhibited the present synthesis. At last, even though we made our SLR approach transparent and parts of the review were conducted by two authors independently, an influence of individual bias cannot be ruled out.

7   Conclusion

In response to the RQ, our descriptive SLR contributes to the existing knowledge base by yielding insights on frequently discussed topic dimensions regarding small businesses participating in DP ecosystems and by pointing at research gaps in IS and small business literature. It emphasizes that there is more than meets the eye in the conjunction of platform and small business research. Hypothetically, the strong focus on DP adoption, digital business models, strategies and DT in this area, all referring to a change that is about to happen, might tell us that the common journey of "traditional" firms and DPs has only just begun.

Furthermore, De Reuver et al.'s [37] observations could partly be confirmed: we find that many DP-case studies allow for limited generalizability, for instance due to lacking uniformity in firm size specifications. Moreover, especially human factors influencing DP-related decisions in small businesses seem to bare greater potential for investigation. We also object that micro businesses, despite being the majority in most economies, have received less attention than larger firms did. At the same time, the term "SME" appeared to be a popular synonym for everything not large. This finding supports OECD [29] reporting a lack of micro firm coverage. The report also states that most data on small firm's DP use stems from e-commerce and social media platforms - our study confirms that there exist stronger focal points regarding both DP types, although B2B platforms receive growing attention, too. On top of that, more research opportunities emerge in the direction of small businesses' sustilience in navigating DP ecosystems, hence, a demand for appropriate DP design knowledge arises. Further research gaps concern responsible and human-centered guidance towards DP use in the wake of ongoing industry disruptions. This goes along with promising topics towards the development of innovation capabilities in small businesses, and business owners' competencies to grasp and lead DP-induced change. Finally, literature reviews have found to be scarce in publication outlets of the given research area. Therefore, we would like to encourage the intensification of knowledge-based research efforts in this field.

References

1. Acs, Z.J., Song, A.K., Szerb, L., Audretsch, D.B., Komlósi, É.: The evolution of the global digital platform economy: 1971–2021. Small Bus. Econ. **57**, 1629–1659 (2021)
2. Eisenmann, T.: Strategies for two-sided markets. Harvard Bus. Rev. **84**(10) (2006)
3. Hein, A., et al.: Digital platform ecosystems. Electron. Mark. **30**, 87–98 (2020)
4. Parker, G.G., van Alstyne, M.W., Choudary, S.P.: Platform revolution: how networked markets are transforming the economy - and how to make them work for you. w. w. norton company (2016)
5. Weill, P., Woerner, S.: What's your digital business model?: six questions to help you build the next-generation enterprise. Harvard Business Review Press (2018)
6. Mintzberg, H.: The strategy concept i: five ps for strategy. Calif. Manage. Rev. **30**, 11–24 (1987)
7. Porter, M.E.: Competitive Advantage: Creating and Sustaining Superior Performance. Free Press (1985)
8. Porter, M.E.: Competitive Strategy: Techniques for Analyzing Industries and Competitors. Free Press (1980)

9. Lewrick, M.: Business Ökosystem Design. Vahlen (2021)
10. Mauborgne, R., Kim, W.C.: Blue Ocean Shift: Beyond Competing - Proven Steps to Inspire Confidence and Seize New Growth. Pan Macmillan UK (2017)
11. Mini, T., Widjaja, T.: Tensions in digital platform business models: a literature review. In: Proceedings of the Fortieth International Conference on Information Systems (ICIS), 6. (2019)
12. Hilbolling, S., Berends, H., Deken, F., Tuertscher, P.: Complementors as connectors: managing open innovation around digital product platforms. R&D Manage. **50**, 18–30 (2020)
13. Gassmann, O., Frankenberger, K., Czik, M.: The Business Model Navigator: 55 Models That Will Revolutionise Your Business. FT Publishing International (2015)
14. Hagiu, A., Wright, J.: Multi-sided platforms. Int. J. Ind. Organ. **43**, 162–174 (2015)
15. Evans, D.S., Schmalensee, R.: Failure to launch: Critical mass in platform businesses. Rev. Netw. Econ. **9**(4) (2010)
16. OECD. SME and Entrepreneurship Outlook 2021. https://www.oecd.org/industry/smes/SME-Outlook-2021-Country-profiles.pdf. Accessed 6 Jan 2023 (2021)
17. Drechsler, A., Hönigsberg, S., Watkowski, L.: What's In an SME? Considerations for Scoping Research on Small and Medium Enterprises and Other Organisations in the is Discipline (2022)
18. USITC: Small and Medium-Sized Enterprises: Overview of Participation in US Exports, Investigation No 332–508, USITC Publication 4125. (2010)
19. Kraus, S., Mahto, R.V., Walsh, S.T.: The importance of literature reviews in small business and entrepreneurship research. J. Small Bus. Manage. **61**(3), 1095–1106 (2021)
20. vom Brocke, J., Simons, A., Niehaves, B., Niehaves, B., Riemer, K., Plattfaut, R., Cleven, A.: Reconstructing the giant: on the importance of rigour in documenting the literature search process. In: Proceedings of the 17th European Conference on Information Systems (ECIS), pp. 2206–2217 (2009)
21. Jöhnk, J., Ollig, P., Oesterle, S., Riedel, L.-N.: The complexity of digital transformation-conceptualizing multiple concurrent initiatives. In: Proceedings of the 15th Internationale Tagung Wirtschaftsinformatik, pp. 1051–1066. (2020)
22. Vial, G.: Understanding digital transformation: a review and a research agenda. J. Strateg. Inf. Syst. **28**, 118–144 (2019)
23. Bouwman, H., Nikou, S., de Reuver, M.: Digitalization, business models, and SMEs: How do business model innovation practices improve performance of digitalizing SMEs? Telecommunic. Policy **43**(9), 101828 (2019)
24. Cenamor, J., Parida, V., Wincent, J.: How entrepreneurial SMEs compete through digital platforms: the roles of digital platform capability, network capability and ambidexterity. J. Bus. Res. **100**, 196–206 (2019)
25. Wan, X., Cenamor, J., Parker, G., Van Alstyne, M.: Unraveling platform strategies: a review from an organizational ambidexterity perspective. Sustainability **9**(5), 734 (2017)
26. Tushman, M.L., O'Reilly, C.A.: Ambidextrous organizations: managing evolutionary and revolutionary change. Calif. Manage. Rev. **38**, 8–29 (1996)
27. Heikkilä, M., Bouwman, H., Heikkilä, J.: From strategic goals to business model innovation paths: an exploratory study. J. Small Bus. Enterp. Dev. **25**, 107–128 (2017)
28. Evans, P.C., Gawer, A.: The Rise of the Platform Enterprise: A Global Survey. The Center for Global Enterprise (2016)
29. OECD: The Digital Transformation of SMEs, OECD Studies on SMEs and Entrepreneurship. OECD Publishing (2021)
30. Garzoni, A., De Turi, I., Secundo, G., Del Vecchio, P.: Fostering digital transformation of SMEs: a four levels approach. Manag. Decis. **58**, 1543–1562 (2020)
31. Lasagni, A.: How can external relationships enhance innovation in SMEs? new evidence for Europe. J. Small Bus. Manage. **50**, 310–339 (2012)

32. Corvello, V., Straffalaci, V., Filice, L.: Small business antifragility: how research and innovation can help survive crises and thrive. Int. J. Entrep. Innov. Manag. **26**, 252–268 (2022)
33. Culkin, N., Smith, D.: An emotional business: a guide to understanding the motivations of small business decision takers. J. Cetacean Res. Manag. **3**, 145–157 (2000)
34. Wang, C., Walker, E.A., Redmond, J.: Explaining the lack of strategic planning in SMEs. the importance of owner motivation. Int. J. Organisational Behav. **12**(1), 1–16 (2007)
35. Franco, M., Matos, P.G.: Leadership styles in SMEs: a mixed-method approach. Int. Entrepreneurship Manage. J. **11**, 425–451 (2015)
36. Franco, M., Prata, M.: Influence of the individual characteristics and personality traits of the founder on the performance of family SMEs. Eur. J. Int. Manag. **13**, 41–68 (2019)
37. De Reuver, M., Sørensen, C., Basole, R.C.: The digital platform: a research agenda. J. Inf. Technol. **33**, 124–135 (2018)
38. Paré, G., Trudel, M.-C., Jaana, M., Kitsiou, S.: Synthesizing information systems knowledge: a typology of literature reviews. Inform. Manage. **52**, 183–199 (2015)
39. Vom Brocke, J., Simons, A., Riemer, K., Niehaves, B., Plattfaut, R., Cleven, A.: Standing on the shoulders of giants: challenges and recommendations of literature search in information systems research. Commun. Assoc. Inform. Syst. **37** (2015)
40. Okoli, C.: A Guide to Conducting a Standalone Systematic Literature Review. Commun. Assoc. Inform. Syst. **37** (2015)
41. Snyder, H.: Literature review as a research methodology: an overview and guidelines. J. Bus. Res. **104**, 333–339 (2019)
42. Kraus, S., Breier, M., Dasí-Rodríguez, S.: The art of crafting a systematic literature review in entrepreneurship research. Int. Entrepreneurship Manage. J. **16**, 1023–1042 (2020)
43. Yoo, Y., Henfridsson, O., Lyytinen, K.: Research commentary: the new organizing logic of digital innovation: an agenda for information systems research. Inf. Syst. Res. **21**, 724–735 (2010)
44. Webster, J., Watson, R.T.: Analyzing the past to prepare for the future: Writing a literature review. MIS Quart. xiii-xxiii (2002)
45. Wirdiyanti, R., Yusgiantoro, I., Sugiarto, A., Harjanti, A.D., Mambea, I.Y., Soekarno, S., Damayanti, S.M.: How does e-commerce adoption impact micro, small, and medium enterprises' performance and financial inclusion? Evidence from Indonesia. Electron. Comm. Res. 1–31 (2022)
46. European Commission: Commission Recommendation of 6 May 2003 concerning the definition of micro, small and medium-sized enterprises (Text with EEA relevance) (notified under document number C(2003) 1422). Eur-Lex. URL: https://eur-lex.europa.eu/legal-content/EN/TXT/?uri=CELEX:32003H0361, Accessed 6 Jan 2023 (2003)
47. Spriggs, M., Yu, A., Deeds, D., Sorenson, R.L.: Too many cooks in the kitchen: innovative capacity, collaborative network orientation, and performance in small family businesses. Fam. Bus. Rev. **26**, 32–50 (2012)
48. Mandviwalla, M., Flanagan, R.: Small business digital transformation in the context of the pandemic. Eur. J. Inf. Syst. **30**, 359–375 (2021)
49. Berendes, I., Zur Heiden, P., Niemann, M., Hoffmeister, B., Becker, J.: Usage of local online platforms in retail: insights from retailers' expectations. In: Proceedings of the 28th European Conference on Information Systems (ECIS), p. 46 (2020)
50. European Commission: Annual Report on European SMEs 2021/2022 - SMEs and environmental sustainability. SME Performance Review 2021/2022 (2022)
51. OECD: Structural business statistics ISIC Rev. 4. (2015)
52. Karjaluoto, H., Huhtamäki, M.: The role of electronic channels in micro-sized brick-and-mortar firms. J. Small Bus. Entrep. **23**, 17–38 (2010)

53. Mkansi, M.: E-business adoption costs and strategies for retail micro businesses. Electron. Commer. Res. (2021). https://doi.org/10.1007/s10660-020-09448-7
54. Bollweg, L., Lackes, R., Siepermann, M., Weber, P.: Drivers and barriers of the digitalization of local owner operated retail outlets. J. Small Bus. Entrep. **32**, 173–201 (2020)
55. Heimburg, V., Wal, N., Wiesche, M.: Professionalizing Small Complementors in a Heterogeneous Platform Ecosystem. A Logistics Case. In: Proceedings of the 17th Internationale Tagung Wirtschaftsinformatik (WI), p. 5 (2021)
56. Benitez, J., Arenas, A., Castillo, A., Esteves, J.: Impact of digital leadership capability on innovation performance: the role of platform digitization capability. Inform. Manage. **59**, 103590 (2022)
57. Burgess, S.: Representing small business web presence content: the web presence pyramid model. Eur. J. Inf. Syst. **25**, 110–130 (2016)
58. Deilen, M., Wiesche, M.: The Role of Complementors in Platform Ecosystems. In: Ahlemann, F., Schütte, R., Stieglitz, S. (eds.) WI 2021. LNISO, vol. 48, pp. 473–488. Springer, Cham (2021). https://doi.org/10.1007/978-3-030-86800-0_33
59. Gierlich-Joas, M., Schüritz, R., Hess, T., Volkwein, M.: SMEs' approaches for digitalization in platform ecosystems SMEs' approaches for digitalization in platform ecosystems. In: Proceedings of the 23rd Pacific Asia Conference on Information Systems (PACIS), p. 190 (2019)
60. Wu, A., Song, D., Liu, Y.: Platform synergy and innovation speed of SMEs: the roles of organizational design and regional environment. J. Bus. Res. **149**, 38–53 (2022)
61. Asadullah, A., Faik, I., Kankanhalli, A.: an exploratory study into the role of multisided platforms in developing the marketing capabilities of SMEs. In: Proceedings of the 24th Pacific Asia Conference on Information Systems (PACIS) (2020)
62. Pan, L., Xiao, Fu., Li, Y.: SME participation in cross-border e-commerce as an entry mode to foreign markets: A driver of innovation or not? Electron. Comm. Res. (2022). https://doi.org/10.1007/s10660-022-09539-7
63. Bärsch, S., Bollweg, L., Lackes, R., Siepermann, M., Weber, P., Wulfhorst, V.: Local shopping platforms - harnessing locational advantages for the digital transformation of local retail outlets: a content analysis. In: Proceedings of the 14th Internationale Tagung Wirtschaftsinformatik (WI) (2019)
64. Ha, S., Kankanhalli, A., Kishan, J.S., Huang, K.-W.: does social media marketing really work for online SMEs?: An Empirical Study. In: Proceedings of the 37th International Conference on Information Systems (ICIS), p. 4 (2016)
65. Omotosho, B.J.: Small scale craft workers and the use of social media platforms for business performance in southwest Nigeria. J. Small Bus. Entrepreneurship, 1–16 (2020)
66. Rauhut, A., Hiller, S., Lasi, H.: Requirements to Enable Platform-based Ecosystems in the Craft Sector. In: Proceedings of the 25th Pacific Asia Conference on Information Systems (PACIS), p. 65 (2021)
67. Hönigsberg, S.: A Platform for Value Co-creation in SME Networks. In: Hofmann, S., Müller, O., Rossi, M. (eds.) DESRIST 2020. LNCS, vol. 12388, pp. 285–296. Springer, Cham (2020). https://doi.org/10.1007/978-3-030-64823-7_26
68. Camposano, J.C., Haghshenas, M., Smolander, K.: Evaluating the value of emerging digital platform ecosystems: lessons from the construction industry. In: Proceedings of the 29th European Conference on Information Systems (ECIS) (2021)
69. Hiller, S., Weber, P., Rust, H., Lasi, H.: Identifying business potentials of additive manufacturing as part of digital value creation in smes - an explorative case study. In: Proceedings of the 53rd Hawaii International Conference on System Sciences (HICSS) (2020)
70. Marheine, C., Pauli, T., Marx, E., Back, A., Matzner, M.: From Suppliers to Complementors: motivational factors for joining industrial internet of things platform ecosystems. In: Proceedings of the 54th Hawaii International Conference on System Sciences (HICSS) (2021)

71. Scuotto, V., Del Giudice, M., Obi Omeihe, K.: SMEs and mass collaborative knowledge management: toward understanding the role of social media networks. Inf. Syst. Manag. **34**, 280–290 (2017)
72. Foster, C., Bentley, C.: Examining ecosystems and infrastructure perspectives of platforms: the case of small tourism service providers in Indonesia and Rwanda. Commun. Assoc. Inform. Syst. **50**(1), 41 (2022)
73. Holland, C.P., Gutiérrez-Leefmans, M.: A Taxonomy of SME E-Commerce platforms derived from a market-level analysis. Int. J. Electron. Commer. **22**, 161–201 (2018)
74. Pfister, P., Lehmann, C.: Digital value creation in German SMEs – a return-on-investment analysis. J. Small Bus. Entrepreneurship, 1–26 (2022)
75. Asadullah, A., Faik, I., Kankanhalli, A.: Can digital platforms help SMEs develop organizational capabilities? a qualitative field study. In: Proceedings of the 41st International Conference on Information Systems (ICIS), p. 12 (2020)
76. Madill, J., Neilson, L.C.: Web site utilization in SME business strategy: the case of Canadian wine SMEs. J. Small Bus. Entrep. **23**, 489–507 (2010)
77. Santoso, A.S., Prijadi, R., Balqiah, T.E.: How open innovation strategy and effectuation within platform ecosystem can foster innovation performance: Evidence from digital multi-sided platform startups. J. Small Bus. Strateg. **30**, 102–126 (2020)
78. Li, X.L., Troutt, M.D., Brandyberry, A., Wang, T.: Decision factors for the adoption and continued use of online direct sales channels among SMEs. J. Assoc. Inf. Syst. **12**(1), 1–31 (2011)
79. Li, L., Su, F., Zhang, W., Mao, J.-Y.: Digital transformation by SME entrepreneurs: a capability perspective. Inf. Syst. J. **28**, 1129–1157 (2018)
80. Bartelheimer, C., Betzing, J.H., Berendes, C.I., Beverungen, D.: Designing multi-sided community platforms for local high street retail. In: Proceedings of the 26th European Conference on Information Systems (ECIS). (2018)
81. Schallmo, D., Williams, C.A., Tidd, J.: the art of holistic digitalisation: a meta-view on strategy, transformation, implementation, and maturity. Int. J. Innov. Manag. **26**, 2240007 (2022)
82. Gartner: What's New in the 2022 Gartner Hype Cycle for Emerging Technologies. URL: https://www.gartner.com/en/articles/what-s-new-in-the-2022-gartner-hype-cycle-for-emerging-technologies, Last Visited 6 Jan 2023 (2022)
83. Weigel, A., Heger, O., Hoffmann, J., Röding, K.: CEOs of SMEs: How IT-governance compensates the lack of digital competencies. In: Proceedings of the 28th European Conference on Information Systems (ECIS) (2020)
84. Grant, J., Wunder, T.: Strategic transformation to : learning from COVID-19. J. Strateg. Manag. **14**, 331–351 (2021)
85. Weiner, B.J.: A theory of organizational readiness for change. Implement. Sci. **4**, 67 (2009)
86. Parker, C.M., Castleman, T.: Small firm e-business adoption: a critical analysis of theory. J. Enterp. Inf. Manag. **22**, 167–182 (2009)
87. Fitz, L.R.G., Scheeg, M., Scheeg, J.: Amplifying human factors in the inquiry of SMEs' needs in digitalization collaborations with external service providers. Proc. Comput. Sci. **200**, 595–601 (2022)
88. United Nations: Transforming Our World: The 2030 Agenda for Sustainable Development. Draft resolution referred to the United Nations summit for the adoption of the post-2015 development agenda by the General Assembly at its sixty-ninth session (2015)
89. Mucha, T., Seppala, T.: Artificial intelligence platforms–a new research agenda for digital platform economy. ETLA Working Papers, 76. The Research Institute of the Finnish Economy (ETLA) (2020)
90. Shneiderman, B.: Human-Centered AI. Oxford University Press (2022)

91. Bell, M.L.: Innovation Capabilities and Directions of Development. Brighton: STEPS Centre (2009)
92. Brown, T.: Design Thinking. Harv. Bus. Rev. **86**, 84 (2008)
93. Lawson, B., Samson, D.: Developing innovation capability in organisations: a dynamic capabilities approach. Int. J. Innov. Manag. **05**, 377–400 (2001)
94. Smorodinskaya, N.V., Russell, M.G., Katukov, D.D., Still, K.: Innovation Ecosystems vs. Innovation Systems in Terms of Collaboration and Co-creation of Value. Proceedings of the 50th Hawaii International Conference on System Sciences (2017)
95. Gill, R.: Change management–or change leadership? J. Chang. Manag. **3**, 307–318 (2002)

Political Polarization in Times of Crisis: Ideological Bias and Emotions of News Coverage of the COVID-19 Pandemic on YouTube

Gautam Kishore Shahi[✉]

University of Duisburg-Essen, Duisburg, Germany
gautam.shahi@uni-due.de

Abstract. News can distribute in many forms across social networks in a fragmented media landscape. Misinformation, such as conspiracy theories, can reach people on social media, leading them into polarized spaces. The consequences of this polarization can be a loss of trust in the media and the perception of news that only reflects one's true or false beliefs. In the context of the COVID-19 pandemic, these beliefs could be that wearing masks is unnecessary. Scientific questions are often politicized and subjected to polarized discussions on social media. Yet, it needs to be clarified to what extent these public health questions are used for political purposes by partisan news channels. This study analyzed 13 ideologically diverse news channels, 11,293 news headlines, and 413,583 user-generated comments on the media coverage concerning COVID-19 on the YouTube platform. For method, wordfish is used to find identity ideology and polarisation and sentiment analysis is used for finding polarization. Results show that textual video information throughout most news channels on YouTube was hardly marked by ideological polarization, except on the Fox News channel. Furthermore, ideological polarization was more pronounced within the comment sections of those news channels and was characterized strongly by negative emotions.

Keywords: Political Polarization · Ideological Bias · Wordfish · Emotion · Sentiment Analysis · COVID-19

1 Introduction

Social media have an essential function in distributing news, as they can reach and influence many people quickly. This observation is often associated with concepts such as echo chambers, fake news, misinformation, or filter bubbles embedded in social media as a potential threat to democracy [1–4]. All these concepts indicate a certain form of dysfunctionality within communication processes concerning issues of public interest. Considering this potential dysfunctionality in online communication, especially in times of the COVID-19 pandemic, there is a pressing need to investigate the behavior of news channels and the extent to

© The Author(s), under exclusive license to Springer Nature Switzerland AG 2023
J. Maślankowski et al. (Eds.): PLAIS EuroSymposium 2023, LNBIP 495, pp. 56–73, 2023.
https://doi.org/10.1007/978-3-031-43590-4_4

which they provide partisan coverage to address concerns about these biases in news coverage [5,6]. Research on political communication in social media has provided evidence for both (a) potential polarization, that is, political opinions becoming more extreme [7,8] and (b) potential depolarization, that is, political opinions becoming more moderate [9–12]. One potential explanation for political polarization through social media is often offered by ideologically charged media coverage in a fragmented media landscape [7], often accompanied by ideologically and emotionally charged user-generated comments [13]. Still, there is not much scientific evidence about the ideological bias in the news coverage on social media. A question that appears of particular relevance in the context of a universal health crisis such as the COVID-19 pandemic.

Especially the effects of biased reporting may lead individuals into so-called "filter bubbles" which can influence the formation of political opinions [14], even if these contain dangerously false information such as "onions or garlic act against the virus and kill it" or "infusion of disinfectants act against the virus". Addressing the idea of political filter bubbles on YouTube, a recent study showed a high level of homogeneity of right-wing populist videos in the recommendation network [15]. In addition to YouTube, a recent study based on Twitter has also found that 15% of tweets during the U.S. presidential election 2016 contained extremely biased news linked to a news medium [16]. Weatherly and colleagues found that the participants perceived headlines from CNN and Fox News differently; specifically, CNN was perceived more liberally than Fox News [17]. Due to the fragmented media landscape and the fact that the news coverage of the same topics is presented in diverse ways, natural language processing (NLP) techniques allow researchers to evaluate how different textual elements are represented on social media platforms. One unsupervised learning method based on quantitative text scaling, which was originally used to analyze the political ideology of party speeches, is Wordfish [18]. Wordfish has been used to identify ideology from political texts [19] and social media [20], as well as to identify polarization from news articles [21]. Although some research has been carried out on the polarization of COVID-19 on news articles [22], no studies have been found that investigate the combination of news that is published by news-providing channels and their corresponding comments on the basis of their political ideology on the topic of COVID-19. To provide a comprehensive insight into the potential ideological bias of news coverage on YouTube, this work proposes an extended NLP approach that examines the ideology of news channels and their news coverage (based on video) in combination with user-generated comments; we first ask:

RQ1: *To what extent have the news media coverage and user-generated comments on YouTube on the COVID-19 topic been ideologically politically polarized?* Given that the news cycle concerning public health issues is volatile, often determined by the emergence of new scientific evidence, a comprehensive analysis of news coverage requires a dynamic approach to mapping the evolution of ideological bias in the news over time.

RQ2: *Are there temporal fluctuations in the level of ideological polarization within the COVID-19-related news coverage and comment sections on YouTube?*

In addition to the ideological polarization associated with news channels, the identification of emotions to the text-based information of videos and comments is also relevant, as previous research has shown that there is a significant influence of the user's online environment on emotional behavior [23].

RQ3: *What emotions and sentiment are within COVID-19-related news coverage and comment sections on YouTube?* To address these three questions, we investigate news channels on YouTube, which are the key players in the continuous dissemination of information about the current information on COVID-19 (e.g., CNN, BBC, and Fox News), and compare the textual elements such as titles, description and tags with each other based on media bias. Using an empirical NLP approach, we measured the polarization of YouTube videos during the COVID-19 pandemic based on various news channels with different news biases. Furthermore, we apply sentiment analysis to the corpus.

2   Background

In this section, we provide an overview of previous work done on polarization and media bias on social media and the identification of ideological polarization and Sentiment.

2.1   Polarization and Media Bias on Social Media

The polarization of political systems is a far-reaching problem that can be observable offline and on social media [24]. Spohr points out that ideological polarization is accompanied by serious consequences, wherein the most concerning consequence "is the loss of diversity of opinions and arguments" [25](p. 151). Especially in the COVID-19 crisis, in which the social consensus on the COVID-19 virus could be key to political decisions, it seems that debates are not always marked by rationality but are often accompanied by extreme opinions and growing doubts, anger and fear: Initial studies on the emotional tone of social media discussions, particularly at the onset of the COVID-19 pandemic, have already revealed that negative emotions, such as fear, were predominantly reflected in opinions [26–28]. Based on a recent Pew Research report, the partisan polarization in the trust in the news has increased in recent five years [29]. This raises the question of how much the media landscape contributes to polarization - at least in systems in which the media landscape is ideologically fragmented (as in the USA; [30]). A recent study has shown that increasing polarization is driven by political misperceptions that can be fostered by media coverage [31]. YouTube has become a popular platform allowing international news outlets to distribute news content and allow individual users to view and comment on this content [32]. In a Pew Research Center study regarding news gathering on YouTube, it was found that 26% of U.S. adults get news from YouTube [33]. A recent study has also shown that many politically controversial topics in YouTube comments are off-topic and thus less likely to take a concrete pro/contra stance [34]. Polarization is a problem, especially in the USA [2,24,35]. It has been observed that many science-related issues, such as COVID-19, are politicized and associated with a polarized tone in media coverage [22,36,37]. One boosting factor for this problem is thought to be the fragmented media landscape and the notion that news channels are often ideologically biased [38]. The area of partisan bias has already been examined in many different studies, which makes it all the more important to compare further research with different analytical methods in order to be able to generate a detailed picture of the media landscape. Still, it is unclear whether the partisan nature of news coverage is also reflected in social media channels and whether it is reconstructed in the user-generated content that is associated with that coverage (e.g., user comments). With this study, we aim to address this question.

2.2   Identification of Ideological Polarization and Sentiment

NLP is the bridge between humans and computers, providing a technique for presenting natural human language to computers in a computational form they can understand. The identification of ideological polarization and sentiment analysis are both aspects that can be further investigated with NLP techniques on social media data. For the identification of the ideological polarization of political texts and news articles, an unsupervised learning algorithm called Wordfish has already been successfully used in several studies [19–21]. Wordfish was introduced as an unsupervised learning technique to identify the ideology of a political party based on the election manifesto [19]. Later, it was applied to several other text sources like social media data and political speech analysis. Wordfish is a statistical model for text scaling. It scales the text based on the word frequency in the given document by assuming the Poisson process generates the word frequencies. The Wordfish algorithm depends on four combined parameters, which are shown in the following equations:

$$word_{ij} \sim Poission_{(\lambda_{ij})}$$

$$\lambda_{ij} = e(\alpha_i + \varphi_j + \beta_j \theta_i)$$

where α is the document lengths, φ is the average word frequencies, β is the word-specific weight and θ estimates the ideological position. $word_{ij}$ is the count of word i in the document j.

Chinn et al. (2020) used the Wordfish modeling to determine the polarization and politicization of climate change news content, and they showed that over the period, involvement of both left and right-wing is highly polarised and politicized. Also, during the beginning of COVID-19, the news was highly polarized and politicized [22]. Aydogan et al. (2019) conducted a study to present the ideological position of social media text and implement the approach for tweets. Following the approach, in this study, we used the textual information from videos and their top comments to find ideological polarization. The political position in the YouTube comment has been analyzed and found. Besides identifying the ideological position of politically relevant texts, detecting the sentiment and its polarity is also important in determining the opinion climate of the news content and the user-generated content comments. To analyze the textual elements of news and user-generated content from comments according to their positive or negative stance, these automated sentiment analyses are ideally suited to investigate the polarity because they compute the semantic connections of words in the text [39]. In this case, a lexicon-based procedure is used, in which negative and positive words with sentiment values in dictionaries are defined in advance and can differ strongly depending on the domain. The final determination of the sentiment of a text from words stringed together is done by weighting the words and calculating the average values of all words. In our research case, we used the NRC emotion lexicon [40]. This lexicon was realized by crowdsourcing on MTurk and contained eight emotions (anger, fear, anticipation, trust, surprise, sadness, joy, and disgust) and two sentiment categories (positive and negative). The NRC emotion lexicon has already been conducted to study sentiment on online social platforms regarding COVID-19 and news headlines [41]. Due to the widespread use of the lexicon, it provides the opportunity to compare the results of our analysis with other studies. Further studies have also shown that this is possible to determine ideological polarization in online social media using network analytic techniques [42, 43]. These studies examining the Twitter platform revealed that political information is more polarized and that the diffusion of information is more homogeneous among like-minded users. Studies on sentiment analysis of online social media refer, on the one hand, purely to the textual elements [44, 45] or otherwise also in combination with a network analysis are carried out [34, 46], in order to be able to make further statements of the users about communication channels and relationship in this respect. Focusing on textual characteristics, the prevalence of automatic sentiment detection in the political context has attracted research interest. Aslam and colleagues examined media coverage on different media sources using sentiment analysis to reveal that more than half (52%) of all news headlines analyzed had negative sentiment (fear, trust, anticipation, sadness, and anger) with negative polarity, while headlines with positive emotions only accounted for 18%; in contrast, 30% of headlines were neutral [44]. Williams et al. (2015) combined sentiment and network analysis on the Twitter platform to analyze online communication on climate change. They found that most people who participate in online discussions are embedded in communities with like-minded users and that these messages usually have a positive mood, while messages between sceptics and activists have a negative mood.

3   Method

This section provides a detailed overview of data collection, finding polarization using swordfish and sentiment analysis.

3.1   Data Collection

For the selection of the YouTube channels regarding their political media bias, we have used the platform "AllSides," a news aggregator that has classified various news channels according to their political bias. The variety of methods used by the platform to determine the political media bias (e.g., blind surveys, academic research) has the advantage that they strengthen the validity of the outcome and contribute to the overall consistency [47]. Furthermore, the platform might serve as a baseline for classifying the political leaning from news articles [48]. According to a recent survey by the Pew Research Center, YouTube is the most used online platform among the U.S. population with 73% [49]. To examine the distribution of news on the subject of COVID-19, we selected YouTube channels with different political orientations. Especially during the COVID-19 pandemic, it is important to estimate how much influence news coverage has and which identifying features these textual elements can have to attract a lot of attention. The dataset collection is based on real data from the YouTube platform, which is collected directly from the YouTube API. We focused on YouTube video information from the channels by calling the object "playlist list" and "video list" from the API using a Python program to collect data [6]; we collected all videos with their metadata from 30.12.2019 to 01.09.2020 and stored them in a MySQL database. In total, we collected 22,508 records of videos. Since our goal is to study the distribution of news from the different news channels with their media bias about the COVID-19 pandemic, we filtered the dataset by the following terms ["corona," "coronavirus," "virus," "covid," "covid-19", "lockdown," "mask," "pandemic," "outbreak"] according to the title, description, and tags. After applying the filter, we have a number of 11,293 videos that are related to COVID-19. Table 1 shows the amount of data collected in relation to the individual message channels and their topic.

Table 1. Display of the YouTube channels with their bias rating from "Allsides"

YouTube Channel	Bias Rating	Videos	Comments
The New York Times	Left	66	3015
CNN	Left	912	54,508
Huffington Post (HuffPost)	Left	262	4163
MSNBC	Lean Left	3249	155,619
POLITICO	Lean Left	102	2308
BBC News	Center	819	41,874
Forbes	Center	169	2337
Reuters	Center	3294	50,705
New York Post	Lean Right	223	1880
The Washington Examiner	Lean Right	177	235
Breitbart News	Right	370	11,139
Daily Caller	Right	249	3792
Fox News	Right	1401	82,008
Total		11,293	413,583

Besides the individual video, the comment area, i.e., what the different users discuss publicly, is also relevant to determine the ideological climate of opinion. Additionally, using this filtered video data set, we have collected the top ten relevant comments and their 10 replies by date using the YouTube API for each video. We did so because we expect these user-generated texts to have generated the most attention from other users while watching the videos. In total, we collected 413,583 comments. To get a better overview and understanding of the data, we conducted an empirical analysis that considers the temporal aspect.

3.2   Wordfish

Before applying the unsupervised learning technique Wordfish, we merged the textual information (i.e., title, description, and tags) of the video data set to represent more meaningful content. Since users come into contact when watching a video with different textual elements such as the title, description, or tags and thus generate their attention, we have decided to summarize them as one unique variable for the following analyses. This means that we have formed a long sentence from the title of the description and the tags to bundle this information. This has the advantage of providing a complete representation of the video where different textual elements come together. For further analysis, we worked with the R package "quanta" [50] to transform the texts into a document term matrix, which is a common process in text processing where the number of terms in the documents is represented in the form of a mathematical matrix. As a further preprocessing step, we removed punctuation and excluded stop words. Studies of NLP have shown that removing stopwords can improve performance for further analyses such as text classification or sentiment analysis [51–53]. Since the tags are used to search for specific keywords for YouTube to suggest videos and find them better, many channel operators have taken the name of the channel in different variations as a tag. For this reason, we also decided for the video record to add them as another stopword and exclude them for further analysis so that the topic of the video can be better represented and no bias arises. Furthermore, we have created a reduced document matrix and thus trimmed the size by setting a threshold of 5, where this number of terms must occur in the document. The final step was to group the document term matrix into months and media bias to prepare it for the Wordfish algorithm, which used a Poisson scaling to estimate document position in one dimension [18]. The Poisson distribution rate depends on four parameters: text length, frequency of the word, the position of the word underlying position, and the weight of the word. The Poisson model assumes that a token frequency in a text is independent of other tokens. With the algorithm, it is easy to scale the political party ideology. To get the information about frequency and informativeness, we can plot a curve between α and β, and plotting θ gives estimated positions of ideal points. We grouped documents by their media bias and their months to consider temporary viewing in the analysis.

3.3   Sentiment Analysis

The same data preprocessing steps were used for the sentiment analysis for the Wordfish algorithm. To measure the sentiment and emotion in the videos as well as in the comments, we used the NRC Emotion Lexicon [40], which allows us to calculate the sentiment of positive and negative words as well as to measure the emotion such as anger, fear, joy, and surprise. The authors of the NRC Emotion Lexicon showed that using the lexicon can lead to statistically significant improvements over the majority baseline classifiers and justify these results with the manual annotation of words for emotions, which requires many resources [54]. Previous studies have also used the NRC Emotion Lexicon for sentiment analysis when examining emotions about the COVID-19 pandemic [44,45,55]. We applied the lexicon also in conjunction with the quanteda package to study the same characteristics of the preprocessed text of the videos and the comments. Here, we grouped the feature frequencies in the document frequency matrix by month and media bias to weight the features according to their relative frequency. This allows a temporal comparison based on the months to the individual news channels and their media bias to specify more precisely how often positive and negative words were used in the articles.

(a)   (b)

Fig. 1. Estimated theta over the whole time period for the different YouTube channels and their media bias based on a) videos and b) comments.

4   Results

Regarding RQ1, Figs. 1a and 1b show the ideological position of the different YouTube channels and their media bias based on videos and comments over the whole investigation period. Previous research has also used Wordfish as a research method to analyze the policy positions from text data, where the determination of the estimate has shown that negative values represent the left ideology, positive values with the right ideology, and values around the zero are expected as centrist [56]. Since we know the ideological position of the channels in advance, we can also assume in our consideration of thetas and adapt the principle to examine the texts for their ideology. The computation of the ideological position of textual information of the videos to the different news channels is mainly moderate, with two exceptions (MSNBC, Fox News). Based on the results of Wordfish, Fox News received a positive theta value of 3.15 which is classified as having a bias toward the right and also disseminating the ideologically strongest right-wing texts in news coverage. MSNBC, in contrast, received a negative theta value of -1.25 which is classified with the media bias lean left, indicating more ideologically leftist textual information. In contrast, the examination of the comments and their computed ideological position showed a stronger variance in comparison to the information in the videos provided by the news channels. For the news channels classified as left and lean left by AllSides, the evaluation of the comments revealed that in all cases except "The New York Times," the theta (representing the extent of ideological polarization) is in a positive range of values, which means that these comments are more likely to follow a rightist ideology. On the other hand, the news channels with a central media bias show a negative theta rating in the comments and can therefore be classified as leftist. On news channels with a right and lean right media bias, there is a tendency for people in these comments to represent the right ideology. The only exception is the New York Post, which has a negative theta value and is, therefore, more likely to be leftist. However, the frequency of comments from lean-right news channels is rather underrepresented compared to the other media bias categories.

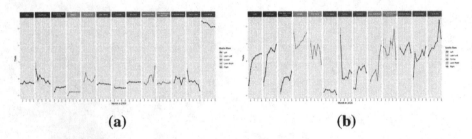

(a)   **(b)**

Fig. 2. Estimated theta per month for the YouTube channels and their media bias based on a) videos and b) comments.

Addressing RQ2, Fig. 2a shows the temporal communication process in YouTube channels and their media bias with the ideological position of the content published from January until the end of August. The results based on the videos indicated that for the YouTube channels (Forbes, BBC, Reuters), the measured theta value remains constant over the different months, but for some channels (The Washington Examiner, Fox News, MSNBC, CNN, New York Times) there are only very short fluctuations of the theta value. In particular, the left-leaning and lean-left channels (HuffPost and POLITICO) show strong fluctuations, especially in the first months from January to March. The New York Post channel tends to show a steady right-oriented increase from March to June until it falls again in July and then rises sharply in August. On the other hand, the results from the Breitbart channel showed that it tends to behave moderately, publishing more left-leaning content in February and June. In the case of the Daily Caller channel, it only starts at a theta value of 0.5 and, over time, finds itself in the negative value range and thus becomes more left-oriented.

The monthly analysis of our results based on the comments in Fig. 2b showed a very strong variance in the level of polarization in the discussion space. Over the months, the values of the ideological position vary greatly depending on

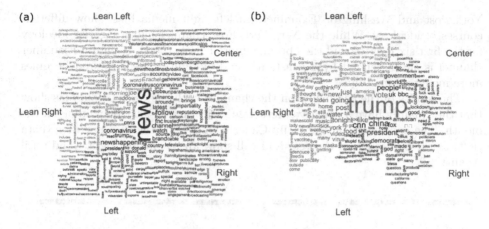

Fig. 3. Word cloud representing the most frequent words clustered by the media bias on a) videos and b) comments.

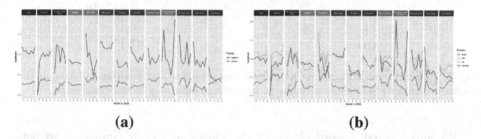

(a)   **(b)**

Fig. 4. a) Sentiment analysis to investigate the negative and positive polarity in the video information, and b) Sentiment analysis to investigate emotions in the video information.

the channel. In general, the comments on the channels with the left media bias show that they had a negative theta value at the beginning of the pandemic (left ideology) and increased steadily over the months (right ideology) until they ended with a positive value at the end of August (CNN, HuffPost). The channel of the New York Times showed a similar course but remained in the negative theta range in August. The two channels, Politico and MSNBC, with a lean left media bias, show similar patterns in the ideology of comments. The channel Politico shows a regular change between moderate and rightist ideologies, while MSNBC starts with a negative theta value in January and becomes a bit of a moderator in February until the end of August when it shows a similar value as in January. The results of the comments on the channels with the center media bias show that it is the most left of all channels over the different months. The comments to the channel Reuters have a negative Theta value in January and February, which turns to zero until August and thus becomes moderate. On the other hand, the comments on Forbes start with a more rightist ideology, although this also becomes more leftist as the months go by. The two channels, New

York Post and Washington Examiner, on left-right media bias, show different courses of ideology. While the New York Post channel had a rightist ideology for the first six months and then became more leftist, the Washington Examiner channel is characterized by a constant trend of rightist ideology and becomes more moderate in August.

The results of the comments on the channels with the right media bias show that they mainly have a negative theta value and vary slightly depending on the month. The comments on the Daily Caller channel start with a negative theta (left) at the beginning of the year and will show an ideological right trend until August.

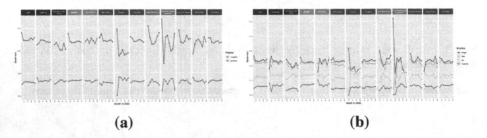

(a)   **(b)**

Fig. 5. a) Sentiment analysis to investigate the negative and positive polarity in the comments, and b) Sentiment analysis to investigate emotions in the comments.

We also created a word cloud for the videos and their comments to better understand the words' textual properties for the individual bias categories. Figure 3a shows that the most common word for video information is news, which is obvious since YouTube channels are news channels. Furthermore, it can be seen from the figure that different variations of the term "coronavirus" exist. It is also noticeable that the channels with the media bias left and lean left use political actors ("Trump," "Donald," and "Joe") more frequently in their video information. The comparison with the word cloud in Fig. 3b shows the words used in the comments of the videos with the respective media bias. It is noticeable that the most frequent word used was "trump." Comments on videos with the left media bias more often use words that are directly related to COVID-19 and its consequences, like "death," "rate," "doctors," "disease," or "school." Furthermore, vulgar words like "stupid," "bullshit," and "shit," or words like "fake" and "lying" can be found. In contrast, comments on videos with the media bias lean left used political terms related to the U.S. presidential election, such as "vote," "election," "republicans," "senate," "america." However, video comments with the media bias center show words related to the government and also use words related to the United Kingdom ("london", "uk", "boris", "british", "britain"), which originated from the BBC YouTube channel. Comments on videos with the lean right media bias also more often use words directly related to the pandemic, such as "vaccine," "virus", "wear", "mask."

The most common words in the comments of the media bias on the right refer to words like "china", "president", "hoax", "american" and "democrats".

To answer RQ3, we present the sentiment and emotional analysis for the YouTube videos and comments. Here we present line charts that visualize the evolution of the polarity (negative and positive) of the video information and comments and the emotions (anger, fear, joy, and surprise) over the investigated timespan. Figure 4a shows the relative frequency of polarity from the video information (title, description, and tags) for each YouTube channel with their specific media bias. The results show that negative polarity dominates for almost all channels throughout the eight months, except for the Fox News channel. The negative polarity drops steadily from January to August while the relative number of positive sentiments increases. In combination with Fig. 4b, which shows the relative frequency of emotion for the textual information of the videos, especially emotions such as anger and fear are most dominant for all news channel's overall media bias. There are also individual news channels, such as Politico, where an increase in the relative frequency of joy can be seen from March to July. Also surprising are the results for the Fox News channel, where anger decreased consistently from January to August 2020 while joy increased continuously.

To also examine the comments and their polarity and emotions, we also conducted a sentiment analysis on this, and the results can be seen in Figs. 5a and 5b. Similar to the results of the polarity of the video information, the results on the comments show a similar result. Again, the relative frequency of comments showing a negative sentiment is more dominant than comments showing a positive sentiment, which shows over the entire study period as well as overall news channels. While the relative frequency of comments tends to be constant over time, there are more frequent fluctuations in the sentiment of the negative polarity. Figure 5b, which deals with the emotions in the comments, illustrates that anger and fear dominate over the entire period and across all news channels. The emotional category surprise was the least common.

5 Discussion

This study investigated the ideological polarization of YouTube videos and comments during the COVID-19 pandemic based on various news channels with different news biases. Our results suggest that most news channels (except Fox News) have published ideologically rather neutral news on COVID-19. In contrast, the analysis of user-generated comments showed an enhanced level of ideological polarization, constantly changing over time [44]. At the same time, they are in line with findings by systematic news analysis indicating an ideologically neutral news coverage [22]. Our finding needs to be interpreted by considering the fact that we have analyzed only the title, description, and tags and not the actual content of the videos, which in fact, contain news and may adopt an ideological attitude. Besides the ideological analysis of textual data to evaluate the media bias in video information, we were also able to determine their sentiment. We found that within the news channels, the meta-information of the video showed a negative sentiment more often than a positive sentiment over the whole period of this investigation. More precisely, emotion, anger, and fear were most dominant from January to August 2020 for each news channel. These findings are in line with the research of Aslam and colleagues, who examined the headlines from 25 English news sources according to their sentiment and found out that 52% of the headlines resulted in a negative sentiment and only 30% in a positive sentiment [57]. While the political tone of the video descriptions was mostly relatively neutral, a more distinct dynamic of the ideology in the comments was observed. This polarization and the resulting dynamic could be an indication of how users perceive the news and how merely biased news coverage can provoke polarized discussions among its viewers and commenters. While the commenters may not represent the group of all viewers, they are relatively impactful in how they may shape other viewers' perceptions of public opinion (leading to subjective estimates of a polarized opinion climate). Research focusing on textual data in online social platforms related to the COVID-19 pandemic has already revealed that predominantly negative emotions, such as fear, were reflected in opinions [26–28]. These insights into negative emotions were also found in our comments, where the proportion of negative sentiment dominated over the entire period of the examination. We were able to clearly identify the emotions, fear, and anger in the temporal observation of the comments. As Duggan and Smith pointed out, this raises the question of the extent to which the general negativity on social media about political issues reflects the political climate of opinion or whether this effect is exclusively reflected in social platforms [58]. However, it must be mentioned that our techniques to analyze sentiment and emotion and the procedure of previous research is based on the NRC lexicon [26,27,44]. By using the lexicon-based method based on words to determine sentiment, there could well be misconceptions since the semantics of the text cannot be considered here. This also means that, for example, sarcastic or ironic content could be associated with the wrong sentiment. As Taboada et al. stated out, the usage of the lexicon-based approach and their included words in the dictionary may not reflect the domain of interest and can therefore provide incorrect results [59]. What are the practical implications of these findings? In times of pandemic, people rely on news and want to keep themselves informed about the latest news of the virus or what actions the government is going to take next. This could be inhibited by content that follows certain political ideologies and shapes emotions, which can have an impact on the climate of opinion. We noticed that the ideological polarization, as well as the emotions in the comments and videos on the YouTube channels, are varying. In terms of practical implications, our findings could be interesting for media outlets to examine their own content for their emotional mood and moderate the comment sections to remain a rationality-based discourse that is less driven by emotions. The results from our study as well as from other studies, could reveal that the existence of negative news about the COVID-19 pandemic shows a high proportion on social platforms. Since previous studies have shown that the pandemic can cause serious health consequences such as depression and anxiety [60,61], future research could examine whether there are links between media news coverage and health consequences. In terms of polarization and political ideology, this could also be helpful for social media users to identify which ideological "footprint" the current video and their comments are in. Moderation of extreme political videos could help create new discussions that are more moderate and do not cover only one political opinion.

6   Limitations

One of the first limitations to be noted is the fact that we did not include all comments in the analysis but focused on the ten most relevant comments and their replies, which may not be completely generalizable. Of course, it seems plausible to assume that those comments with a high level of emotionality are those with more replies. Thereby, we assumed that these comments often received a high number of likes, which might cause them to have an effect on other viewers and thus polarize the video. Furthermore, our analysis focused only on text-based information about the videos. Future research could explicitly address and investigate the video content or the associated subtitles by extracting the transcripts of the videos and analyzing the whole video frame with NLP approaches. They could help to examine a detailed picture of the message content inside the videos by the YouTube channels. Another limitation is that we restricted our analysis to English-language news channels only and thus cannot make any further cross-national statements about the polarization of other news channels on YouTube.

7   Conclusion and Outlook

The present research offers preliminary evidence that the news coverage of the investigated YouTube channels with different media biases is mostly ideologically neutral. Still, there are some exceptions (MSNBC, Fox News) in whose channels the textual information of the videos showed a clear ideological tendency in favor of the already existing media bias. With the present paper, we have performed an analysis based on the connection between medium (video) and users' comments which have a major role in the perception of news, the sentiment of information, and the formation of public opinion, especially when the climate of opinion develops in a polarized ideological direction. The results regarding the polarization of the comment sections show that the political ideology of the various YouTube channels fluctuated greatly over the months. Still, there were some channels where comments tended to develop in a direction where the trend was towards the rightist ideology (CNN, HuffPost, New York Times, Daily Caller, Fox News). Channels with the media bias "center" also had fluctuations in a few months but were generally more leftist. Our results also indicated that negative emotions and sentiments dominated not only the descriptive information of the videos but also the comments.

References

1. Mitchell, A., Gottfried, J., Stocking, G., Walker, M., Fedeli, S.: Many Americans say made-up news is a critical problem that needs to be fixed. Pew Res. Center **5**, 2019 (2019)
2. Sunstein, CR.: # Republic: Divided democracy in the age of social media. Princeton University Press (2018)
3. Shahi, G.K., Dirkson, A., Majchrzak, T.A.: An exploratory study of covid-19 misinformation on twitter. Online Social Netw. Media **22**, 100104 (2021)
4. Shahi, G.K., Nandini, D.: Fakecovid-a multilingual cross-domain fact check news dataset for covid-19. arXiv preprint arXiv:2006.11343 (2020)
5. Shahi, G.K., Majchrzak, T.A.: Amused: an annotation framework of multimodal social media data. In: Intelligent Technologies and Applications: 4th International Conference, INTAP 2021, Grimstad, Norway, October 11–13, 2021, Revised Selected Papers, pp. 287–299. Springer (2022). https://doi.org/10.1007/978-3-031-10525-8_23
6. Röchert, D., Shahi, G.K., Neubaum, G., Ross, B. and Stieglitz, S.: The networked context of covid-19 misinformation: informational homogeneity on youtube at the beginning of the pandemic. Online Social Netw. Media **26**, 100164 (2021)
7. Asker, D., Dinas, E.: Thinking fast and furious: emotional intensity and opinion polarization in online media. Public Opin. Q. **83**(3), 487–509 (2019)
8. Schmidt, A.L., et al.: Anatomy of news consumption on facebook. Proc. National Acad. Sci. **114**(12), 3035–3039 (2017)
9. Bail, C.A., et al.: Exposure to opposing views on social media can increase political polarization. Proc. National Acad. Sci. **115**(37), 9216–9221 (2018)
10. Beam, M.A., Hutchens, M.J., Hmielowski, J.D.: Facebook news and (de) polarization: Reinforcing spirals in the 2016 us election. Inform. Commun. Society, **21**(7), 940–958 (2018)
11. Lee, J., Choi, Y.: Effects of network heterogeneity on social media on opinion polarization among south koreans: focusing on fear and political orientation. Int. Commun. Gaz. **82**(2), 119–139 (2020)
12. Lee, J.K., Choi, J., Kim, C., Kim, Y.: Social media, network heterogeneity, and opinion polarization. J. Commun. **64**(4), 702–722 (2014)
13. Kwon, K.H., Cho, D.: Swearing effects on citizen-to-citizen commenting online: a large-scale exploration of political versus nonpolitical online news sites. Social Sci. Comput. Review, **35**(1), 84–102 (2017)
14. Pariser, E.: The filter bubble: What the Internet is hiding from you. penguin UK (2011)
15. Röchert, D., Weitzel, M., Ross, B.: The homogeneity of right-wing populist and radical content in youtube recommendations. In: International Conference on Social Media and Society, pp. 245–254 (2020)
16. Bovet, A., Makse, H.A.: Influence of fake news in twitter during the 2016 US presidential election. Nature Commun. **10**(1), 7 (2019)
17. Weatherly, J.N., Petros, T.V., Christopherson, K.M., Haugen, E.N.: Perceptions of political bias in the headlines of two major news organizations. Harvard Int. J. Press/Politics **12**(2), 91–104 (2007)
18. Slapin, J.B., Proksch, S.O.: A scaling model for estimating time-series party positions from texts. Am. J. Political Sci. **52**(3), 705–722 (2008)
19. Proksch, S.O., Slapin, J.B.: How to avoid pitfalls in statistical analysis of political texts: The case of germany. German Politics, **18**(3), 323–344 (2009)

20. Ceron, A., Curini, L., Iacus, S.M., Porro, G.: Every tweet counts? how sentiment analysis of social media can improve our knowledge of citizens' political preferences with an application to Italy and France. New Media Society **16**(2), 340–358 (2014)
21. Chinn, S., Hart, P.S., Soroka, S.: Politicization and polarization in climate change news content, 1985–2017. Sci. Commun. **42**(1), 112–129 (2020)
22. Hart, P.S., Chinn, S., Soroka, S.: Politicization and polarization in Covid-19 news coverage. Science Commun. **42**(5):679–697 (2020)
23. Del Vicario, M., et al.: The spreading of misinformation online. Proc. National Acad. Sci. **113**(3), 554–559 (2016)
24. Fletcher, R., Cornia, A., Nielsen, R.K.: How polarized are online and offline news audiences? a comparative analysis of twelve countries. Int. J. Press/Politics **25**(2), 169–195 (2020)
25. Spohr, D.: Fake news and ideological polarization: filter bubbles and selective exposure on social media. Bus. Inf. Rev. **34**(3), 150–160 (2017)
26. Lwin, M.O., et al.: Global sentiments surrounding the covid-19 pandemic on twitter: analysis of twitter trends. JMIR Public Health and Surveillance **6**(2), e19447 (2020)
27. Medford, R.J., Saleh, S.N., Sumarsono, A., Perl, T.M., Lehmann, C.U.: An "infodemic": leveraging high-volume twitter data to understand early public sentiment for the coronavirus disease 2019 outbreak. In: Open forum infectious diseases, volume 7, pp. ofaa258. Oxford University Press US (2020)
28. Xue, J., Chen, J., Chen, C., Zheng, C., Li, S., Zhu, T.: Public discourse and sentiment during the Covid 19 pandemic: Using latent Dirichlet allocation for topic modeling on twitter. PLoS ONE **15**(9), e0239441 (2020)
29. Jurkowitz, M., Mitchell, A., Shearer, E., Walker, M.: Us media polarization and the 2020 election: a nation divided (2020)
30. Arceneaux, K., Johnson, M., Murphy, C.: Polarized political communication, oppositional media hostility, and selective exposure. J. Politics **74**(1), 174–186 (2012)
31. Wilson, A.E., Parker, V.A., Feinberg, M.: Polarization in the contemporary political and media landscape. Curr. Opinion Behav. Sci. **34**, 223–228 (2020)
32. al Nashmi, E., North, M., Bloom, T., Cleary, J.: Promoting a global brand: A study of international news organisations' youtube channels. J. Int. Commun. **23**(2), 165–185 (2017)
33. Stocking, G., Van Kessel, P., Barthel, M., Matsa, K.E., Khuzam, M.: Many Americans get news on youtube, where news organizations and independent producers thrive side by side (2020)
34. Röchert, D., Neubaum, G., Ross, B., Brachten, F., Stieglitz, S.: Opinion-based homogeneity on youtube: combining sentiment and social network analysis. Comput. Commun. Res. **2**(1), 81–108 (2020)
35. Liu, S., Guo, L., Mays, K., Betke, M., Wijaya, D.T.: Detecting frames in news headlines and its application to analyzing news framing trends surrounding us gun violence. In: Proceedings of the 23rd Conference on Computational Natural Language Learning (CoNLL), pp. 504–514 (2019)
36. Shahi, G.K., Clausen, S., Stieglitz, S.: Who shapes crisis communication on twitter? an analysis of german influencers during the COVID-19 pandemic. In: 55th Hawaii International Conference on System Sciences, HICSS 2022, Virtual Event / Maui, Hawaii, USA, January 4–7, 2022, pages 1–10. ScholarSpace (2022)
37. Shahi, G.K., Kana Tsoplefack, W.: Mitigating harmful content on social media using an interactive user interface. In: Social Informatics: 13th International Conference, SocInfo 2022, Glasgow, UK, October 19–21, 2022, Proceedings, pp. 490–505. Springer (2022). https://doi.org/10.1007/978-3-031-19097-1_34

38. Prior, M.: Media and political polarization. Annu. Rev. Polit. Sci. **16**, 101–127 (2013)
39. urney, P.D., Littman, M.L.: Measuring praise and criticism: inference of semantic orientation from association. ACM Trans. Inform. Syst. (tois), **21**(4), 315–346 (2003)
40. Mohammad, S.M., Turney, P.D.: Crowdsourcing a word-emotion association lexicon. Comput. Intell. **29**(3), 436–465 (2013)
41. Khoo, C.S.G., Johnkhan, S.B.: Lexicon-based sentiment analysis: comparative evaluation of six sentiment lexicons. J. Inf. Sci. **44**(4), 491–511 (2018)
42. Barberá, P., Jost, J.T., Nagler, J., Tucker, J.A., Bonneau, R.: Tweeting from left to right: is online political communication more than an echo chamber? Psychol. Sci. **26**(10), 1531–1542 (2015)
43. Boutyline, A., Willer, R.: The social structure of political echo chambers: variation in ideological homophily in online networks. Polit. Psychol. **38**(3), 551–569 (2017)
44. Aslam, F., Awan, T.M., Syed, J.H., Kashif, A., Parveen, M.: Sentiments and emotions evoked by news headlines of coronavirus disease (covid-19) outbreak. Hum. Social Sci. Commun. **7**(1) (2020)
45. Sharma, S., Sharma, A.: Twitter sentiment analysis during unlock period of covid-19. In: 2020 Sixth International Conference on Parallel, Distributed and Grid Computing (PDGC), pp. 221–224. IEEE (2020)
46. Williams, H.T., McMurray, J.R., Kurz, T., Lambert, F.H.: Network analysis reveals open forums and echo chambers in social media discussions of climate change. Global Environ. Change **32**, 126–138 (2015)
47. Chen, W.F., Wachsmuth, H., Al Khatib, K., Stein, B.: Learning to flip the bias of news headlines. In: Proceedings of the 11th International Conference on Natural Language Generation, pp. 79–88 (2018
48. Hamborg, F., Donnay, K., Gipp, B.: Automated identification of media bias in news articles: an interdisciplinary literature review. Int. J. Digit. Libr. **20**(4), 391–415 (2019)
49. Perrin, A., Anderson, M.: Share of us adults using social media, including facebook, is mostly unchanged since 2018 (2019)
50. Benoit, K., et al.: quanteda: an R package for the quantitative analysis of textual data. J. Open Source Softw. **3**(30), 774–774 (2018)
51. Saif, H., Fernandez, M., Alani, H.: Automatic stopword generation using contextual semantics for sentiment analysis of twitter. In: CEUR Workshop Proceedings, vol. 1272 (2014)
52. Saif, H., Fernandez, M., He, Y., Alani, H.: On stopwords, filtering and data sparsity for sentiment analysis of twitter (2014)
53. Shehu, H.A., Tokat, S., Sharif, M.H., Uyaver, S.: Sentiment analysis of turkish twitter data. In: AIP Conference Proceedings, volume 2183, p. 080004. AIP Publishing LLC (2019)
54. Mohammad, S.M., Kiritchenko, S.: Using hashtags to capture fine emotion categories from tweets. Comput. Intell. **31**(2), 301–326 (2015)
55. Dubey, A.D., Tripathi, S.: Analysing the sentiments towards work-from-home experience during Covid-19 pandemic. J. Innov. Manage. **8**(1), 13–19 (2020)
56. Lo, J., Proksch, S.O., Slapin, J.B.: Ideological clarity in multiparty competition: a new measure and test using election manifestos. British J. Political Sci. **46**(3), 591–610 (2016)
57. Budak, C., Goel, S., Rao, J.M.: Fair and balanced? quantifying media bias through crowdsourced content analysis. Public Opinion Quart. **80**(S1), 250–271 (2016)

58. Duggan, M., Smith, A.: The tone of social media discussions around politics. Pew Research Center. http://www.pewinternet.org/2016/10/25/the-tone-of-social-media-discussions-around-politics (2016)
59. Taboada, M., Brooke, J., Tofiloski, M., Voll, K., Stede, M.: Lexicon-based methods for sentiment analysis. Comput. Linguist. **37**(2), 267–307 (2011)
60. Mazza, M.G., et al.: Anxiety and depression in covid-19 survivors: Role of inflammatory and clinical predictors. Brain Behav. Immun. **89**, 594–600 (2020)
61. Nina Vindegaard and Michael Eriksen Benros: Covid-19 pandemic and mental health consequences: systematic review of the current evidence. Brain Behav. Immun. **89**, 531–542 (2020)

Education 3.0 – AI and Gamification Tools for Increasing Student Engagement and Knowledge Retention

Catalin Vrabie(✉) (iD)

National University of Political Studies and Public Administration, Bucharest, Romania
catalin.vrabie@snspa.ro

Abstract. The rapid advancements in Web 2.0 applications and artificial intelligence (AI) have significantly influenced the educational domain, introducing novel challenges for both educators and learners in online learning contexts. By continually learning and adopting innovative teaching methods, educators can cultivate enhanced learning experiences and collaboration between teachers and students. Effective strategies for delivering course content in online environments can further contribute to students' satisfaction and in-depth understanding of the subject matter. This article aims to present a framework for educators to improve interaction with students, grounded in pertinent e-learning literature and supported by institutional research as it discusses the results of a pilot project designed to stimulate learning by tailoring it to students' individual needs and maximizing the advantages of visual technologies in the learning process through gamification. Six courses on various subjects were developed, but this article focuses solely on a computer science course. The outcomes indicate visible improvements in student academic performance and satisfaction and that is to be seen over a comparison on two groups of students – with the second group beneficiating by a new way of delivering course materials.

Keywords: e-learning · e-interaction · Artificial Intelligence

1 Introduction

In recent years, the educational landscape has undergone a significant transformation, driven primarily by advancements in technology. As we enter deeper in the era digitalization, the integration of artificial intelligence (AI) and gamification tools has emerged as a promising approach for enhancing student engagement and knowledge retention. In this paper, we delve into the potential of AI and gamification as catalysts for shaping the future of education, by examining their applications, effectiveness, and the challenges associated with their implementation. Through a rich analysis of existing literature and institutional research conducted by the author, the paper aims to provide valuable insights into how these emerging technologies can revolutionize the learning process, create immersive educational experiences, and facilitate the development of skills necessary for Net Generation students of the 21st century [1]. By contextualizing the findings within

© The Author(s), under exclusive license to Springer Nature Switzerland AG 2023
J. Maślankowski et al. (Eds.): PLAIS EuroSymposium 2023, LNBIP 495, pp. 74–87, 2023.
https://doi.org/10.1007/978-3-031-43590-4_5

the broader discourse on educational innovation, this paper contributes to the ongoing dialogue on the role of technology in creating transformative learning experiences and shaping the future of education.

While AI is thought to have the potential to revolutionize any aspect of life [2–4], modern students expect a more innovative approach to education. The growing demand has led to a surge in investment in educational infrastructure, with an emphasis on developing advanced and adaptable learning resources.

The research question from which the author started in writing this article is, knowing the potential value of AI and gamification and whether there are any benefits of applying AI and gamification to education system at tertiary level? In order to answer this, the author will firstly present AI and gamification tools in education as they are found in the scientific literature proceeding forward into institutional research. Subsequently, the author will be able to extract the benefits and propose AI and gamification solutions for educational purposes (nonetheless, it's not limited to this alone [5, 6]). There might be discussions for each of them, mostly because these technologies are just beginning to emerge in the education field, and some have not yet been fully implemented. However, their functionalities, technical, and social benefits have been tested in a pilot project as will be presented throughout the paper. The article does not intend to confirm the effectiveness of the proposed solutions but only to provide them to readers (some could be policymakers) in order to enlarge their views and, perhaps, start incorporating them into strategic development plans for universities.

2 Literature Review

The notion of Life 3.0, as described by Max Tegmark in his book "Life 3.0: Being Human in the Age of Artificial Intelligence" [3], represents an emerging paradigm that envisions the incorporation of cutting-edge technologies like AI into our daily lives. Analogously, Education 3.0 integrates state-of-the-art technologies into learning settings to facilitate a more immersive and efficacious educational experience. Rather than having a fixed definition, Education 3.0 serves as a framework that characterizes a transition in pedagogical methodologies propelled by technological advancements and innovation.

However, there is a growing body of scientific evidence supporting the efficacy of various elements within what this article proposes as being Education 3.0. For instance, several studies have demonstrated the positive effects of AI-driven adaptive learning systems on student out-comes, engagement, and knowledge retention [7–12]. The cited article discusses how intelligent tutoring systems can provide personalized instruction, feedback, and scaffolding, leading to improved student outcomes, engagement, and knowledge retention tailoring the learning experience to individual needs, thereby improving the overall quality of education.

As back as 2015, Brosser described the challenges in e-learning quality development also providing an innovative process management for building effective e-learning course materials [13].

Käser et al. examines the use of Bayesian networks to model and predict student performance in an adaptive learning environment. The authors found that the AI-driven system improved students' learning outcomes and provided valuable insights into their knowledge state [14].

Additionally, research on gamification has shown that incorporating game elements into the learning process can increase motivation, engagement, and even aca-demic performance [15, 16]. By making learning more enjoyable and interactive, gamification can enhance students' intrinsic motivation to learn and help them acquire knowledge more effectively.

While Yıldırım [16] synthesizes the results of 34 studies on the effects of gamification on academic achievement, cognitive load, and motivation finding that gamification has a positive impact on academic achievement and motivation, Khaldi et al., by a systematic review, examines 33 empirical studies focused on the implementation of gamification in higher education settings [17]. Also, in this study the results suggested that gamification interventions positively affect motivation, engagement, and academic performance, although the magnitude of these effects varies depending on the specific gamification elements used.

Based on a literary review and a desk research method, Krumova outlines technologies perspectives for higher education both for didactical approaches as well as for the universities management [18]. Similarly, Salehi & Largani, by using descriptive-analytical research method conducted a research of Iran's e-learning trends and technological implications in higher-education [19]. The findings indicated that the lack of innovation is to be seen as weakness both in the area of universities management as well as in the pedagogical one.

After reviewing the research outlined in Smart Learning Environments (issues 2020–2023), Interactive Learning Environments (issues 2020–2023), Computers in Human Behavior (issues 2020–2023) and Smart Cities and Regional Development (issues 2020–2023), one can conclude that much of the focus is placed on e-learning systems in general and little on platforms that are providing a gamification environment enriched with AI tools.

The reviewed articles helped drawing a picture of what are the current trends in research into the field of education 3.0 which, as a holistic concept, might not have direct empirical evidence. However, the individual components that constitute this framework have been studied and the findings offer support for the potential benefits of integrating advanced technologies and innovative pedagogical approaches in shaping the future of education [20]. However, it is essential to acknowledge that more research is needed to better understand the long-term impact and the most effective strategies for implementing the principles of Education 3.0.

3   Research Methods and Context

The COVID-19 pandemic has led to significant alterations in various aspects of our lives, encompassing the manner in which university courses were conducted and the methods of course material delivery to students [21].

For this study, the author employed two distinct approaches to delivering course materials: (1) traditional e-learning methods involving electronic documents (e.g., docs, pdfs, etc.) made available on a Learning Management System platform such as Moodle [22–24], and (2) an eLearning Authoring Tool that incorporates gamification and AI techniques to maintain student engagement, specifically Livresq [25] – a SCORM compliant software [26]. The aim of the study is to examine the differences in students' grades obtained after the completion of semester exams, comparing the outcomes of the two distinct approaches to course material delivery.

The National University of Political Studies and Public Administration (SNSPA) in Bucharest, Romania, which served as the foundation for this study, experienced two complete academic years (2020/2021 and 2021/2022) of online classes (in addition to the somewhat chaotic second semester of 2020). Throughout the 2020/2021 academic year, the learning environment employed was Moodle, which facilitated asynchronous delivery of course materials, and was supplemented by Webex for synchronous meetings in accordance with the university schedule. Concurrently, in an effort to enhance students' learning experiences during the 2020/2021 academic activities, members of the academic community initiated a project titled "Digital environments for improving the quality of education [...] maximizing the benefits of visual technologies in the learning process through virtual teaching environments" which was funded by the Romanian Ministry of Education.

The named project implied fully transformation of six courses (from bachelor and master programs) by a modern, AI tool, developed by a Romanian software company, aiming to "provide a learning stimulation system, tailored to the individual study needs of students, by creating an informational platform to support and facilitate educational activities and enhance their quality, as well as developing mechanisms, tools, and procedures that align SNSPA's educational offerings with current labor market trends and the international academic community".

In the following academic year (2021/2022), all six courses were fully implemented. Nevertheless, for the purposes of this article, the author will focus solely on one course ("IT fundamentals for Public Administration" designed for first year, bachelor) and its outcomes, as detailed in the subsequent sections.

At this point, the author would like to emphasize that the exams were technically identical (utilizing Moodle's capability to provide summative tests) and relied on the same question bank consisting of 200 questions (100 for laboratory tests and 100 for the theoretical component). From this pool, 60 randomized closed questions/exercises as presented in Fig. 1 (30 questions for the practical exam and same amount for the theoretical one), were selected for both student groups to be presented during the exams (each question/exercise carried a weight of 0.33% in the final grade, and it was given one minute to answer per question/exercise). Another crucial aspect for this study is that the number of students in each group is roughly equivalent, with approximately 300 students in each group and consisted from the entire population of students in the first year of study, bachelor, for both academic years (2020/2021 and 2021/2022) therefore no sampling was needed.

The initial consideration was whether there is any relationship between the time taken to answer questions or complete exercises and the grades achieved. This is crucial to determine if students utilized extra time to search for correct answers online (giving that the exams were given online, so it was difficult to supervise the high number of participants), despite the limited time allocated for each question discouraging such a strategy. Results of the Pearson correlation indicated that there is a significant very small negative relationship between time and grade for both laboratory exam ($r(308) = .189$,

$p < .001$ for group 1 and $r(274) = .162$, $p = .007$ for group 2) and the theoretical one ($r(304) = .118$, $p = .039$ for group 1 and $r(274) = .0691$, $p = .252$ for group 2) for both groups of students, which means the students were answering by their own.

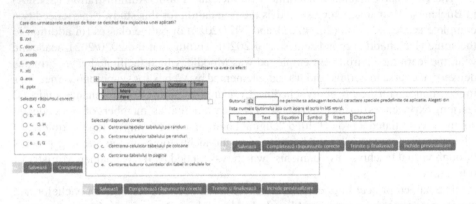

Fig. 1. Different examples of questions addressed during the exam. (Source: apcampus.ro [22])

The Facility index (F) for a randomly generated text, as produced and computed by Moodle, along with the Discrimination efficiency for the same text, is depicted in the following Fig. 2.

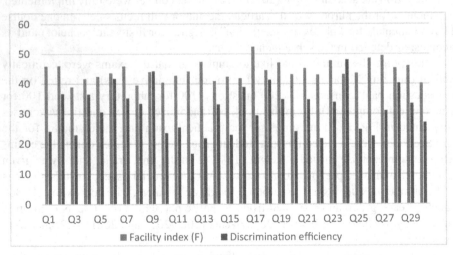

Fig. 2. Question statistics (Data source: apcampus.ro [22])

Discrimination efficiency indicates how effective a question is at sorting out able students from those who are less able – in other words, a question which is very easy or very difficult cannot discriminate between students of different ability, because most of them get the same score on that question [27], the Facility index (F) is giving information about the difficulty of a specific question [27] (comparing with answering time and accuracy registered overall on the entire group of students). As observed in Fig. 2, both indicators suggest that the tests were about right for the average student (evidenced by F values ranging from 35% to 65%) while simultaneously offering a good discrimination index relative to the difficulty of the question (as it is compared with the overall accuracy of the answers given to it) with values between 20%–40% interval – a satisfactory discrimination.

The above information aims to validate the exam's suitability for both student groups. In the following section, we will present detailed results for each group.

4 Results

4.1 First Group Results

The initial group of students, belonging to the 2020/2021 academic year, consisted of 308 individuals. Their scores varied from 0 (implying no correct responses to any question or exercise) to 100 (denoting all questions and exercises were answered accurately). This scoring methodology was applied to both the laboratory and theoretical examinations.

Nonetheless, as the official grading system, in accordance with the regulations, must have values ranging from 1 to 10, we transformed the absolute scores obtained by each student into relative scores by employing the Excel CEILING function in conjunction with the subsequent mathematical formula:

$$Grade_i = \frac{Si - minE}{\frac{maxE - minE}{10}} \tag{1}$$

Were:

Si-Score obtained by student i;

E-Range of scores for that particular exam (practical or theoretical).

In order to avoid a grade such as 0 (zero) we use the following Excel formula to convert absolute scores into relative ones on a scale of 1 to 10:

IF(Di="","",MIN(MAX(CEILING((Di-MIN(D:D))/((MAX(D:D)-MIN(D:D))/10),1),1),10))

$$\tag{2}$$

Were:

Di-is the cell containing the score received by student i (Si as in Formula 1);

D:D-is the column containing the range of scores for that particular exam (E as in Formula 1).

The resulting value is the student's grade given as in the official records, normalized between 1 and 10.

The chart displayed in Fig. 3 illustrates the distribution of grades for both practical and theoretical exams. The vertical axis represents the total count of students who achieved the grade indicated on the horizontal axis. From a statistical perspective, the distribution aligns with expectations. Although the grades may not have been particularly high, over half of the students passed both exams (with an average lab exam grade of 6.40 and a theoretical exam grade of 6.07 – considering only grades higher than 5.00) – Table 1.

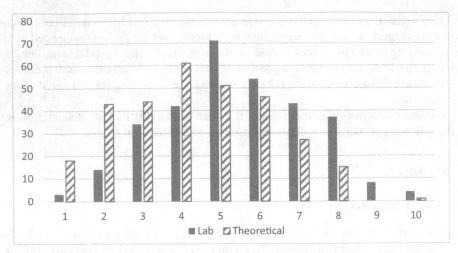

Fig. 3. Distribution of grades received by the first group of students (Data source: apcampus.ro [22])

4.2 Second Group Results

The second group of students, who were enrolled in the 2021/2022 academic year, underwent testing with the same set of questions, as previously mentioned. These questions were similarly randomized by Moodle, and we employed the same approach to derive their grades from their scores, as demonstrated in formulas 1 and 2 in the preceding section.

Upon examining Fig. 4 and Table 1, it is evident that there is a trend towards higher grades, with an average lab exam grade of 6.89 and a theoretical exam score of 7.14, while again taking into account only grades above 5.00.

4.3 General Results

To provide an overview on the aforementioned shift, we present the statistical data for both student groups below, Fig. 5.

Official final grades, as recorded in the academic transcripts, are determined by averaging the laboratory test grade (Lab – represented by the full columns) and the theoretical one (Th – represented by the striped columns), provided that both scores are equal to or greater than 5.00. To compute this, we employed the subsequent Excel formula:

$$= IF(Ai >= 5,IF(Bi >= 5,ROUNDUP(AVERAGE(Ai,Bi),0),Bi),Ai) \qquad (3)$$

Were:

Ai-represents the cell containing the laboratory exam grade for student i (calculated using formula 2);

Bi-represents the cell containing the theoretical exam grade for student i (calculated using formula 2).

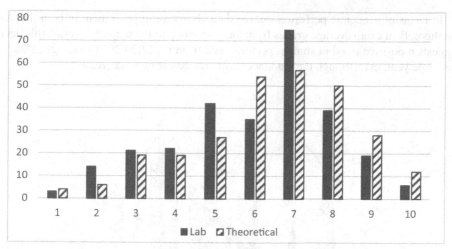

Fig. 4. Distribution of grades received by the second group of students (Data source: apcampus.ro [22])

By overlapping the outcomes for both student groups and incorporating the final grade calculated through the aforementioned formula (blue and red lines), it becomes simpler to observe the shift towards higher grades. It is evident that there was a trend to decrease the grades below 5 (represented by the blue line and bars – Av), while the red bars (and also the red line) become increasingly noticeable in the right portion of the chart, corresponding to higher grades.

Considering that the exam structure for both groups was consistent, the logical interpretation of the data suggests that the second group of students experienced advantages from the newly developed course materials, which were created as part of the specified program. These new course materials have been designed with a focus on enhancing students' engagement while incorporating more effective teaching strategies and utilizing updated techniques to deliver the content. As a result, the students in the second group were able to benefit from these improvements, which likely contributed to their higher performance in the exams.

Table 1. General results data and values.

	First group	Second group
Academic year	2020/2021	2021/2022
No. of students	308	274
Lab exam average grade	6.40	6.89
Theoretical exam average grade	6.07	7.14

Table 1 (above) provides a comparison of academic performance and student numbers between the two groups taken into consideration for this study.

In terms of academic performance, the data shows an improvement in both the lab and theoretical exam average grades from the first group to the second. Overall, this data suggests a positive trend in student performance from the 2020/2021 to the 2021/2022 academic year, even though the student cohort size has slightly decreased.

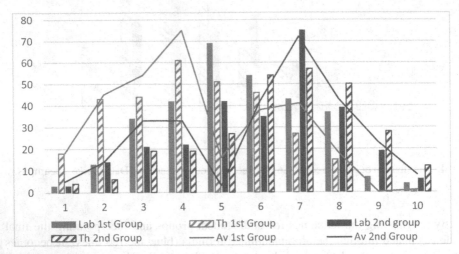

Fig. 5. Distribution of grades received by both groups of students (Data source: apcampus.ro [22])

This demonstrates the importance of continuously updating and refining educational materials to better suit the needs of students and to stay up-to-date with the latest advancements in the e-learning technologies. By doing so, educators can help ensure that their students are well-equipped to excel in their studies and achieve better learning outcomes.

5   The Techniques Employed in Delivering the Course Content for the Second Group of Students

In our project, we incorporated gamification to enhance the learning environment for students. This article aims to provide a scientific examination of the impact of gamification, combined with AI tools, on student achievement through relevant statistical data derived from experimental research. The research compares student performance in two consecutive academic years – the first year (2020/2021) following a traditional e-learning approach, and the second year (2021/2022) employing a more up-to-date method, as described below.

As mentioned, we utilized Livresq, a web-based application software that enables users to create interactive courses and lessons directly from their browser using different gamification techniques in conjunction with Kahoot. The version used in this study can be accessed through the official webpage of the Smart-EDU Hub research group [28].

Upon accessing the course, "the game" starts. Learners are encouraged to engage with the content by opening various sections through clickable buttons they need to discover and read the content. If a user does not interact within ten minutes, the platform closes and records their progress. Also, if the learner does not open all the hidden layers by pressing or scrolling on all the texts or images, he or she won't be allowed to step forward to the nest section.

Moreover, the author of the course materials could record himself audio or video (or he/she could embed a YouTube video in the course material) and, by doing this, he or she can ask students to listen/watch it till the end.

Each chapter concludes with a brief test – like a final quest for advancing in a computer game level (comprising in five to ten different type questions such as: drag and drop into text, drag and drop markers, drag and drop onto image, multiple choice, true/false, select missing words, etc.) designed to review the covered course material. Students must achieve a minimum grade to proceed, although multiple attempts are permitted (unlike exams).

Each step made along the course by the student is somehow unique– the sequence of buttons and clicks differ from one section to another, therefore he or she is forced to stay engaged till the end of each section/chapter. All its time and activity are recorded on the platform and offers a glimpse of the student engagement throughout the semester.

At the time this study, Ascendia, the company providing Livresq, is actively working on incorporating Natural Language Processing (NLP) techniques through collaborations with Microsoft and OpenAI. This integration aims to enable students to engage in conversations with the platform about topics derived from the course materials. In essence, the chatbot will be knowledgeable about the content and will respond to queries accordingly. If a question necessitates a more detailed answer, the chatbot will access the internet to deliver a comprehensive response. By doing so, it further enhances the learning experience for students, as they can receive instant clarification and support for their inquiries, resulting in a more engaging and interactive educational environment. However, this future development horizon is yet in its beta versions since the model needs to adapt to the specific age needs (age identified here by the course material that was accessed by the learner) since the platform is open not only to academics but also to high school gymnasium and even primary school teachers. This aspect is considered very important since the level of understanding both of specific terminology as well as the complexity of a chatbot answer could create confusion.

6 Limitations

Taking into account the significance of this section, we have divided it into two subsections: (1) limitations related to the application of AI and gamification techniques in learning environments, and (2) limitations specific to our own study.

6.1 Weakness of Using AI and Gamification Techniques in e-learning

Although gamification and AI (specifically NLP) techniques provide numerous benefits in the realm of digital education, they also present certain challenges, as outlined below. It is worth mentioning that the author's observations stem from personal experiences while interacting with students and the platform during the course development process and after its deployment. The project involved monthly meetings over a period of eight months both with academics and students to effectively create the courses and gain a deeper understanding of the students' requirements.

- Implementation costs: Developing and integrating gamification and NLP features into e-learning platforms can be expensive and time consuming, as it requires investment in software development, infrastructure, maintenance and subscriptions to dedicated NLP platforms.
- Technology readiness: Not all institutions, professors or students may have access to the necessary technology, stable internet connections, or even digital literacy to effectively utilize e-learning platforms featuring these technologies.
- Overemphasis on rewards: Gamification can sometimes lead to an overemphasis on extrinsic rewards like points, badges, or leaderboards, which might overshadow the intrinsic motivation for learning.
- Potential for distraction: If not designed carefully, the gaming elements or NLP-based chatbots could divert learners' focus from the educational content, leading to suboptimal learning outcomes.
- Scalability issues: As the number of users increases, e-learning platforms incorporating gamification and NLP might face technical challenges in scaling up their systems (and costs as mentioned above), which could result in performance issues or downtime.
- Inaccurate chatbot responses: Despite advancements in NLP, chatbots can still provide incorrect or irrelevant answers to complex questions (a situation that tend to happen with top performance students), which could lead to confusion or misinformation for those in question.
- Personalization limitations: Although gamification and NLP techniques can offer some degree of personalization, they may not fully accommodate the diverse learning styles, preferences, or abilities of all students.

Despite these limitations however, gamification and NLP techniques have the potential to significantly enhance e-learning experiences. By addressing these challenges and continuously refining the implementation of these technologies, educators can create more engaging and effective learning environments for their students.

6.2   Study Limitation

Fully capturing the outcome of "successful student" using a single quantifiable attribute is impossible because of the complex nature of the concept. Regarding our study however, despite the fact that we took into consideration all the possible criteria to properly compare both groups of students, there are few limitations of the analyses that needs to be addressed.

- External factors: Students' performance may be affected by external factors, such as personal circumstances, which can make it challenging to establish a proper comparison of their grades. Nevertheless, the large student sample size in our study allowed for more favorable statistical outcomes.

- Student demographics: Different cohorts of students may have diverse backgrounds, abilities, and learning styles, which can influence their academic performance and make it difficult to accurately compare their grades. This limitation was also overcome due to the large number of students involved in the study.
- Confounding variables: Various factors, such as a better understanding of the environment, can influence students' grades. For example, the first group of students made it clear that learning was the only way to pass an exam, encouraging the second group to focus more on studying. This makes it challenging to isolate the specific causes behind changes in academic performance.

We have addressed the aforementioned limitations to offer a clear view over the comparison itself. There may be additional factors to consider; however, it is evident that Education 3.0 will incorporate AI tools and gamification techniques, gaining increasing momentum in the coming years.

7 Discussions and Conclusions

We consider the terms of digitization and digitalization as related but not synonyms. While the first one only covers the idea of translating analog inputs into digits such as scanning a document into a pdf or jpg, or converting an analog audio recording into an mp3 and making the users a passive actor in its interaction with the information (being only capable to read/watch/listen), the second one refers to the broader transformation of processes and activities to fully leverage the opportunities provided by digital technologies. By digitalization users are allowed to actively interact with the content. Moreover, AI techniques allows machines to convert both text and images into vectors therefore translate one into another (by vectorizing each feature) [29, 30]. What we are witnessing is astonishing an opens a whole new era of development and education needs to keep up since the most important actors in the field were once students so their knowledge and vision is deeply anchored into their educational background.

There are numerous factors to consider as we transition into the era of Education 3.0 – some of which have been discussed in the limitations section, while others remain to be uncovered. However, given the broad range of decisions (from unquestioningly embracing new technologies even if they don't perfectly align with a specific society or community, to outright rejecting them due to imperfections) and the limited time available to make these decisions, institutions (public or private) that rapidly adapt will gain momentum and attract more students. Regrettably, apart from the most dedicated individuals, many enrolled students are not entirely committed to their studies but instead seek the most convenient path [31–33], waiting for new opportunities to emerge. Universities must enter this new era in the same time, providing students with a comprehensive array of choices rather than allowing decisions based solely on external factors.

In our research, we conducted a literature review on gamification elements and the application of AI in digital higher education pedagogical processes, their integration, and the various gamification strategies proposed to modernize learning systems. Drawing on this knowledge and the value gained from our pilot project, we investigated whether exposing the students of the Net Generation to a modern e-learning platform would enhance their overall performance. The study demonstrates that, by utilizing modern tools, students can retain more information and effectively apply it as needed.

References

1. Vrabie, C.: "Book Review: 'Grown Up Digital: How the Net Generation is Changing Your World' by Don Tapscott." HOLISTICA J. Bus. Public Administ. **6**(1), (2015)
2. A. Ng, Interviewee, AI is the New Electricity. [Interview]. 9 11 2018
3. Tegmark, M.: Life 3.0: Being Human in the Age of Artificial Intelligence, Penguin books (2017)
4. Vrabie, C., Dumitrascu, E.: Smart Cities de la idee la implementare. Universul Academic, Bucharest (2018)
5. Vrabie, C.: Artificial intelligence promises to public organizations and smart cities. In: Digital Transformation: 14th PLAIS EuroSymposium on Digital Transformation, PLAIS EuroSymposium 2022, Sopot, Poland, December 15, 2022, Proceedings, Gdansk (2022)
6. Vrabie, C.: E-Government 3.0: An AI model to use for enhanced local democracies. Sustainability **15**(2), 9572 (2023)
7. Ahmad, S., Rahmat, M., Mubarik, M., Alam Hyder, S.: Artificial intelligence and its role in education. Sustainability **13**(22) (2021)
8. Subhash, S., Cudney, E.A.: Gamified learning in higher education: a systematic review of the literature. Comput. Hum. Behav. **87**, 192–206 (2018)
9. Murtaza, M., Ahmed, Y., Shamsi, J.A., Sherwani, F., Usman, M.: AI-based personalized e-learning systems: issues, challenges, and solutions. IEEE Access, pp. 81323–81342 (2022)
10. Kalogiannakis, M., Papadakis, S., Zourmpakis, A.-I.: Gamification in Science Education. A Systematic Review of the Literature. Educ. Sci. **11**(1), (2021)
11. Tahiru, F.: AI in education: a systematic literature review. J. Cases Inform. Technol. **23**(1), (2021)
12. Medium.com, "How Intelligent Tutoring Systems are Changing Education," Medium.com, 1 8 2020. https://medium.com/@roybirobot/how-intelligent-tutoring-systems-are-changing-education-d60327e54dfb. Accessed 2 May 2023
13. Brosser, L., Vrabie, C.: The quality initiative of E-learning in germany (QEG)-management for quality and standards in E-Learning. In: 5th World Conference on Learning, Teaching and Educational Leadership, WCLTA 2014 (2015)
14. Käser, T., Klingler, S., Schwing, A.G., Gross, M.: Beyond knowledge tracing. modeling skill topologies with bayesian networks. In: 12th International Conference, ITS 2014, Honolulu, HI, USA, June 5–9, 2014 (2014)
15. Sailer, M., Hense, J.U., Mayr, S.K., Mandl, H.: How gamification motivates: an experimental study of the effects of specific game design elements on psychological need satisfaction. Comput. Hum. Behav. **69**, 371–380 (2017)
16. Yıldırım, I., Şen, S.: The effects of gamification on students' academic achievement: a meta-analysis study. Interact. Learn. Environ. **29**(8), (2021)
17. Khaldi, A., Bouzidi, R., Nader, F.: Gamification of e-learning in higher education: a systematic literature review. Smart Learn. Environ. **10**, (2023)
18. Krumova, M.: Open data benchmarking for higher education: management and technology perspectives. Smart Cities Regional Develop. (SCRD) J. **1**(2), pp. 47–60 (2017)
19. Salehi, S.J., Largani, S.M.H.: E-Learning challenges in Iran's higher education system and its implications in the realm of good governance. Smart Cities and Regional Development (SCRD) J. **4**(1), 9–21 (2020)

20. Vrabie, C.: Education – a key concept for E-administration. Procedia. Soc. Behav. Sci. **186**, 371–375 (2015)
21. Iancu, D., Vrabie, C., Ungureanu, M.: Is blended learning here to stay? public administration education in Romania. In: Central and Eastern European eDem and eGov Days, Budapest (2021)
22. Faculty of Public Administration, SNSPA, "apcampus.ro," Faculty of Public Administration, SNSPA, https://apcampus.ro/. Accessed 2 May 2023
23. Vrabie, C.: ELEMENTE DE E-GUVERNARE [Elements of e-government]. Pro Universitaria, Bucharest (2016)
24. Vrabie, C.: Just do it–spreading use of digital services. In: EGPA Conference, Malta (2009)
25. Ascendia, "LIVRESQ: eLearning Authoring Tool," Ascendia https://livresq.com/en/. Accessed 2 May 2023
26. SCORM.com, "SCORM solved and explained," https://scorm.com/. Accessed 2 May 2023
27. Moodle, "Quiz report statistics," moodle, https://docs.moodle.org/dev/Quiz_report_statistics. Accessed 2 May 2023
28. Smart-EDU Hub, "e-QUAL EDU," Smart-EDU Hub, 2023. https://www.smart-edu-hub.eu/events/projects. Accessed 2 May 2023
29. Radford, A., Jozefowicz, R., Sutskever, I.: Learning to Generate Reviews and Discovering Sentiment (2017)
30. Ouyang, L., et al.: Training language models to follow instructions with human feedback. Adv. Neural. Inform. Process. Syst. **35**, 27730–27744 (2022)
31. Hu, M., Li, H.: Student engagement in online learning: a review. In: 2017 International Symposium on Educational Technology (ISET), Hong Kong (2017)
32. Kahu, E.R.: Framing student engagement in higher education. Stud. High. Educ. **38**(5), 758–773 (2013)
33. Kizilcec, R.F., Mar, P.-S., Maldonado, J.J.: Self-regulated learning strategies predict learner behavior and goal attainment in massive open online courses. Comput. Educ. **104**, 18–33 (2017)

Privacy Pattern Catalog Approach for GDPR Compliant Appliance: From Legal Requirements to Technology Design

Lukas Waidelich(✉) and Thomas Schuster

Pforzheim University, Tiefenbronner Str. 65, 75175 Pforzheim, Germany
lukas.waidelich@hs-pforzheim.de

Abstract. Digital services must adapt to new developments in technology, policy, and compliance regulations. Data protection plays a critical role, making the implementation of privacy-aware services essential. However, small and medium-sized enterprises struggle with meeting data protection requirements. In this paper, we address the challenges posed by the GDPR and compare them with existing solutions. Additionally, we discuss mechanisms to translate legal requirements into technological design. Our approach utilizes patterns as foundational guides to assist in the efficient implementation and application of data protection measures. This article introduces four patterns expanding our collection of data protection patterns. Privacy patterns facilitate compliance with regulations during development and operation. Creating a pattern catalog for data protection is the next logical step. We outline an approach for its development.

Keywords: General Data Protection Regulation (GDPR) · EU GDPR · Pattern Catalog · Privacy Pattern

1 Introduction

The core of today's fourth industrial revolution is defined by the generation, exchange, and processing of digital and sensitive data. This focus on data is foundational to our contemporary knowledge society and has led to a considerable increase in the volume of data stored worldwide [1]. A clear distinction is made between personal and non-personal data, both of which may contain sensitive information requiring protection. Concurrently, the past decade has seen the rise and expansion of the data economy, fueled by technological trends such as the Internet of Things (IoT) and Artificial Intelligence (AI). These advancements are part of a broader digital transformation that is disrupting traditional practices, spawning innovative business models, enhancing process efficiency, and generating new products and services. However, despite these technological strides, small and medium-sized enterprises (SMEs) frequently encounter difficulties in implementing the necessary measures to comply with various country-specific regulations, such as the European General Data Protection Regulation (EU GDPR), which govern the protection of personal data. The challenges faced by these enterprises highlight a

J. Maślankowski et al. (Eds.): PLAIS EuroSymposium 2023, LNBIP 495, pp. 88–102, 2023.
https://doi.org/10.1007/978-3-031-43590-4_6

wider issue: legal norms and their enforcement are struggling to keep pace with the rapid technological developments, thereby creating a pressing need for more robust and adaptable legal frameworks to ensure data protection and privacy [2–5].

With this article, we intend to contribute to the field of applied data protection. Through ongoing research activities [6, 7], we are summarizing and developing specific patterns that transform complex data protection requirements into actionable technical or organizational directives, referred to as GDPR patterns. These patterns are being gathered into an inclusive catalog, aiming to serve as a practical guide for those in the field. This GDPR pattern catalog is not only a tool but a means to elevate the comprehension and endorsement of fundamental data protection principles. Our research leads to three significant research questions (RQ):

- **RQ1**: What obstacles are encountered by small and medium-sized enterprises (SMEs) as they strive to implement and manage the GDPR?
- **RQ2**: What existing solutions are in place, and to what extent do they address the challenges identified in RQ1?
- **RQ3**: How might a catalog containing GDPR patterns be thoughtfully constructed and utilized to overcome the challenges pinpointed in RQ1?

These questions form the core inquiry of our research, guiding the exploration of challenges and solutions in the field of applied data protection. They reveal an intricate relationship between the needs of SMEs, existing solutions, and innovative approaches, all centered around the effective use and understanding of GDPR.

This paper is structured as follows: Sect. 2 provides an overview of patterns and EU GDPR. Section 3 explains the scientific methodology and presents the research findings, including the challenges of EU GDPR compliance in companies, the analysis of related works, and the design and development of a GDPR pattern catalog. The paper concludes with an outlook on future research activities and legislative developments.

2 Knowledge

2.1 Pattern Knowledge

The first pattern ideas reach back to the 1970s. The architect Alexander developed 253 design patterns of cities, buildings, and structures. The fundamental idea is to document recurring solution principles in pattern forms. According to Alexander, a pattern consists of a (recurring) problem, a description of the problem with an illustration, and a related solution. Patterns are answers to design problems [8]. The advantages are externalization, structuring and documentation of solutions, reduction of complexity and effort, and creation of uniform communication. The pattern design is used today in a wide variety of disciplines such as software development [9]. Privacy patterns are an interesting area of research. However, these only focus on general privacy approaches [10–12] or specific requirements [6, 7, 13] such as privacy by design. A broad understanding and implementation approach in enterprises is missing.

2.2 Patterns in EU GDPR Context

Suited EU GDPR [14] areas for pattern development are: *GDPR Principles, Rights of the Data Subject, Rights and Obligations for Controllers and Processors* as well as the *Rights of Transfers of Personal Data.* The principles be found in articles 4 to 11. The principles relating to processing of personal data are listed can be found in Article 5. These includes transparency and traceability, purpose limitation, data minimization, accuracy, storage limitation, integrity, and confidentiality, as well as accountability. Further Articles describe the lawfulness of processing and conditions for consent. Articles 12 to 23 of the GDPR describe the rights of the data subject. Important Rights of the Data Subject are information obligation, right of access by the data subject, right to rectification, right to erasure, right to restriction of processing, notification obligation, right to data portability, right to object. Rights and obligations for controllers and processors are described in Articles 23 to 43. Important influences are the regulations regarding privacy by design, privacy by default, security of processing, notification of a personal data breach, data protection impact assessment and data protection officer. The transfer of personal data to third countries or international organizations is covered by Articles 44 to 50. Major points of interest are adequacy decision, appropriate safeguards, binding corporate rules, legal assistance treaty and specific situations. The remaining sections are partially suitable for pattern creation.

3 Research Methodology

This section describes the systematic research approach. We refer to the Design Science Research (DSR) approach of Hevner et al. [15, 16] which is common in the field of Information Systems (IS). The methodological process forms the basis for the conception of design artifacts. The research is guided by six phases proposed by Peffers et al. [17] for the development of DSR in IS (see Fig. 1). The aim is to develop new design artifacts in the form of GDPR patterns. This design knowledge is used to create a GDPR patterns catalog as another design artifact.

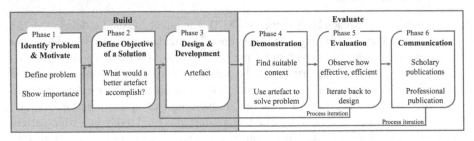

Fig. 1. Procedure model for developing a GDPR pattern catalog [17].

The procedure model used (see Fig. 1) classifies six phases into two fields: Build and evaluate. Our research follows a similar path. In this publication, we address the first three phases (highlighted in gray): (1) Identify Problem & Motivate, (2) Define Objective of a Solution, and (3) Design & Development. The other phases (4) Demonstration, (5) Evaluation, and (6) Communication are currently in progress. Figure 1 illustrates the phases and describes the involving DSR tasks to create the design artifact. A detailed description of phases one to three is given below.

3.1 Problem Identification and Motivation

This phase deals with the problem description and its motivation to tackle it. The specific research challenge should be described. Further, the value of the solution for a target group should be presented. A clear representation of the problem complexity and its delimitation to adjacent research questions is prerequisite for the artifact's development. In this context, the problems might be conceptually divided into different sets of problems. Problem description serves two purposes: First, it engages the researcher community and allows the proposed way of solving the problem. Second, it creates transparency and empathy for further reasoning. Domain knowledge is crucial in understanding the problem's state and the importance of its resolution [17].

3.2 Objectives of a Solution

The second phase describes the deviation between the targeted solution and the problem definition. The objectives can be formulated quantitatively or qualitatively. An illustrative quantitative criterion would be the question, to what extent the new solution is better than the current one. An exemplary qualitative criterion might address the question, what a new solution contributes to a problem that has not been covered. The objectives are derived from the problem specification. Analogous to the first phase, appropriate domain knowledge is required. This contains the problem status and the effectiveness of current solutions (if available) [17].

3.3 Design and Development

The third phase specifies the artifact development. Artifacts refer to entities or combinations of constructs, models, methods, or instantiations [15]. The phase involves defining the desired artifact functionality and specifications, followed by its development with the required design knowledge for solution finding.

4 Results

4.1 Problem Identification and Motivation

In this section the problem-centered approach is explored. This includes a description of the challenges in adopting privacy regulation, specifically the EU GDPR. The difficulties have been documented in the literature for several years. In our daily work with SMEs, we can confirm these challenges. To identify the problems, we have conducted a literature review and synthesized the most important findings. Search for papers contain the two keywords "GDPR" and either "challenge" or "problem". The target group is focused on SMEs and organizations in the European area. The problems are considered in general on an organizational and processual level. The literature under consideration primarily focuses on the period after the implementation of the EU GDPR in May 2018.

Raising data protection awareness in SMEs and organizations is crucial, including awareness of the EU GDPR. Studies conducted after the introduction of the EU GDPR highlighted significant shortcomings in awareness, with parts of the workforce remaining uninformed about the regulations [18] and studies indicate that GDPR awareness continues to be a challenge even today [19]. The evolving regulations and technological advances emphasize the importance of privacy as a relevant cross-sectional topic that demands attention.

Small companies and organizations hardly have access to highly specific expertise related to data protection and legal issues. They have neither resources (expertise) nor human resources (experience). This shortcoming can be summarized as missing knowledge [18, 20–22]. This task is often outsourced and causes high costs, or the task is not prosecuted due to shortened budget [4].

In addition to lacking awareness and expertise, missing training is often mentioned in this context [18]. Training also has to reflect regular or annual changes [23]. Training and knowledge creation can only take place once the organization has achieved awareness. Technical implementation or organizational adaptation of data protection policies throughout the organization often is a major problem [18]. Implementation steps are typically associated with large temporal and monetary efforts. Furthermore, literature claims that there is no standardized procedure. Each company must adapt individually to its context (business models, processes and operations) [23, 24].

Another survey reveals that about half of small businesses have not yet complied with the implementation of the GDPR in multiple respects [2]. Poritskiy et al. explicitly mention three challenges in implementation: Usability, interface management and technical requirements. The implemented solutions must be structured in such a way that the workforce is able to use them. Data protection should already be considered in development (privacy by design). Technical concepts such as anonymization or encryption must be mastered [25].

Regulatory legislation suffers from image problems. This applies especially to privacy. Data protection regulations are perceived as boring, bureaucratic, and administratively complicated. Public opinion towards EU GDPR is viewed negatively. Discussions of this topic often taking place with a negative bias. The other side of the coin is often neglected. Regulation is an important vehicle for government to achieve policy agendas. Regulation sets uniform guidelines and makes it possible to achieve better results for the economy, the environment and society. The GDPR strengthens citizens privacy and creates a uniform regulation in the European area [5].

The review identifies shortcomings despite the European GDPR being in effect for several years. SME and organizations have room for improvement in the areas of awareness, knowledge, training, implementation, and adaptation. From the political side, it can be noted that the EU GDPR faces an image problem.

The rapid growth of this area is intriguing and inspiring from a scientific perspective. Extensive research has been conducted on the root causes of problems and the identification and analysis of challenges. The logical next step is to move from problem descriptions to potential solutions in the research field. It is crucial to determine how data protection challenges can be implemented and applied, with a focus on both organizational measures and technical approaches. For instance, technologies like process mining have integrated automated privacy mechanisms during runtime [25].

4.2 Objectives of a Solution

In this section, we delve into the exploration of a target-oriented solution, presenting existing methods and concepts found in literature. Our inquiry began with a literature review conducted over the past ten years, using key terms like "GDPR," "pattern," and "privacy" to ensure the relevance of our findings. Our findings, which encompass a diverse range of approaches, are carefully detailed in chronological order. Importantly, we also took the step of connecting the identified research findings to previously pinpointed problems, emphasizing existing gaps and shortcomings. This process not only allowed us to understand the landscape of current solutions but also laid the foundation for the creation of our GDPR pattern catalog. The insights we've gleaned are instrumental in shaping this catalog, furthering our aim to provide practical and tailored solutions for data protection.

Hoepman [10] details a data protection strategy to achieve a certain level of privacy. This is based on the preliminary draft of the EU GDPR. Privacy (design) patterns are mentioned as an abstract concept for software developers, which are implemented by privacy enhanced technology. He mentioned eight privacy design strategies: Minimize, hide, separate, aggregate, inform, control, enforce and demonstrate. In a further work [11], an additional abstracting level between the privacy strategy and the privacy design patterns was introduced with tactics. From a hierarchical perspective it can be stated that strategies describe a desired privacy target. Tactics support a higher-level strategy. Patterns formalize an abstract implementation of a tactic. Privacy enhanced technology implements these privacy patterns [26]. Along with other researchers they created a privacy pattern catalog for software developers (refer to www.privacypatterns.org). The catalog currently contains 72 available privacy patterns that support privacy strategies and privacy tactics [27]. In addition, a collection of privacy patterns has been developed (refer to https://privacybydesign.digital). The catalog contains a total of 79 privacy patterns in German language. These patterns have been derived from previous research.

Notario et al. [12] present a way to perform privacy analysis via PRIPARE (PReparing Industry to Privacy-by-design by supporting its Application in REsearch) method. For this purpose, two contrasting approaches (goal-based and risk-based approaches) are used. The PRIPARE method aims to involve as many stakeholders as possible during the software development cycle. This is achieved by a flexible execution of the seven-step process model. The researcher group plans to integrate privacy principles into a general software development tool later.

Huth's publication [28] is the first to point out the research gap considering privacy holistically. Previous work considered merely infrastructure elements and the application and database level. He proposes an interesting approach for his dissertation: The creation of a GDPR pattern catalog based on the (legal) requirements, stakeholders, and solutions. Overarching publications from Huth can be found, but the realization of the GDPR pattern catalog is still missing.

Kung and Martin [13] highlight the significance of methods and tools for GDPR compliance in software engineering. They align with the privacy by design principle, which is a core aspect of the EU GDPR. In their subsequent project, PDP4E (accessible at https://www.pdp4e-project.eu/), they developed software tools for GDPR compliance. These tools encompass risk management, engineering requirements, privacy-aware design, and assurance management. The applications were tested in domains such as connected vehicles and big data on smart grids.

Brodin [23] introduces a framework for SME to comply with GDPR adaptation. His work follows a three-step approach consisting of analysis, design, and implementation. The processes are described in general and cover twelve practical implications for organizations [29]. However, explicit usage instructions (e.g., implementation manual or checklist) could not be found.

In their paper, Roesch et al. [7] present nine privacy patterns tailored to the EU GDPR. These patterns follow a standardized structure, beginning with reference to the GDPR legislation. They then provide a non-legal description of the associated challenge and propose a technical solution. Each pattern concludes with a checklist. The presented patterns were evaluated within a research project, but no transfer measures were mentioned.

The software tool PADRES (PrivAcy, Data REgulation and Security) was developed by researchers Pereira et al. [30]. PADRES is a tool that analyzes web applications and aids in ensuring compliance with data protection regulations. It incorporates the principles of the EU GDPR through a checklist and questionnaire. The tool generates a summary report that includes recommendations based on the analysis. The authors also discuss potential extensions of PADRES, specifically in vulnerability and cookie analysis.

Our research identifies promising concepts, approaches, frameworks, and solutions in privacy enhancing technologies for EU GDPR. Some of these approaches were developed before the introduction of the EU GDPR, while others fully address its requirements. Table 1 maps the identified research works to the challenges faced by SMEs (refer to Sect. 4.1).

Table 1. Overview of identified research findings on EU-GDPR challenges for SMEs.

Research							
Challenges	[10]	[12]	[28]	[13]	[23]	[7]	[30]
Awareness	◐	◐	●	●	●	●	◐
Knowledge	◐	◐	○	●	◐	◐	◐
Training	○	○	○	○	○	◐	◐
Implementation	●	●	◐	●	◐	◐	●
Adoption	○	○	◐	○	●	○	○
○ = no conformity ◐ = partial conformity ● = conformity							

The overview provides insights into the characteristics of the identified research projects. It is notable that most projects aim to raise awareness about GDPR. While most of the works offer comprehensive knowledge about the EU GDPR, they lack aspects of knowledge transfer, such as training formats or guidelines. There is a strong emphasis on technical implementation, targeting individuals with a technical background, such as software developers. However, these approaches may not effectively reach non-technical individuals who deal with GDPR issues daily. As a result, achieving sustainable awareness and expertise among the workforce remains a challenge. Only a minority of the works address the adaptation of GDPR into organizational operations. Recent research has increasingly focused on the needs of SMEs. However, within the scope of the research, a holistic approach that combines organizational adaptation with technical solutions could not be identified.

4.3 Design and Development

We are introducing a proposal for a GDPR pattern catalog, along with the associated GDPR patterns, as a design artifact. This introduction draws on the insights gathered from Sects. 4.1 and 4.2. Initially, we present and elucidate our concept for the GDPR pattern catalog, sharing ideas about a systematic approach to GDPR patterns. Next, we detail the method behind the design of patterns. Finally, we showcase four GDPR patterns that we developed, extending the scope of existing work.

Privacy Patterns Catalog
Usually, patterns are not independent entities. They are combined into a catalog of patterns, forming an extensive library that systematically divides them into several divisions. The next stage of development is a pattern system. A pattern system formalizes the relationships between patterns, their implementation, and the process of finding solutions.

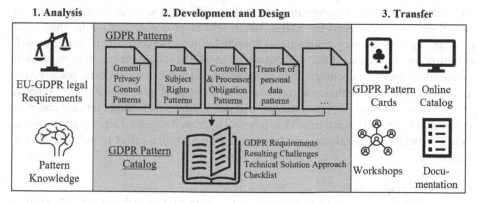

Fig. 2. Overview of the GDPR Pattern System.

In this publication, we provide an overview of our pattern system approach. To illustrate, we will refer to Fig. 2. We divide the pattern system into three phases: 1.

Analysis, 2. Development and Design, and 3. Transfer. The proposed pattern catalog, which falls under the second phase, is at the core of this publication. We also describe initial transfer measures (part of Phase 3).

In the analysis phase we combined two knowledge components: Process knowledge about the legal requirements of the EU GDPR and method knowledge about the development of (software) patterns. It is important to divide the EU GDPR into relevant sections. In doing so, we followed the EU GDPR areas outlined in Sect. 2.2. Four GDPR areas were identified: *General Privacy Control Patterns*, *Data Subject Rights Patterns*, *Controller and Processor Obligation Patterns*, and *Patterns for the Transfer of Personal Data*.

The second phase: Development and design focused on evolving the design artifacts and corresponding pattern catalog. Approximately 30 potential GDPR patterns were identified across these areas. While some patterns have been published [6, 7], others are still being elaborated or developed. The GDPR pattern catalog contains fundamental GDPR knowledge, addressing the findings from Sect. 4.1. In addition, we plan an influence matrix that will be provided to showcase the dependencies between the patterns, and a filter option will be added to the pattern catalog to help stakeholders find relevant patterns. Based on feedback, the catalog may indicate the level of implementation complexity.

As part of the subsequent privacy pattern system development, four measures for the third transfer phase are proposed: We begin by suggesting the development of GDPR pattern cards for practical application with stakeholders. These cards, illustrated in Fig. 3, feature a template with various color schemes related to specific GDPR areas (as explained in Sect. 2.2). Each card will include information on GDPR requirements, challenges, stakeholders, fields of application, implementation complexity, and dependencies, with technical solution approaches and checklists on the reverse side, all available online. Next, we focus on the GDPR catalog, designed as an online tool to facilitate effective utilization. This catalog offers advanced features such as filters and cross-linkages between patterns, allowing a comprehensive understanding of their interdependencies, and complexity levels to gauge implementation difficulty. We then turn our attention to collaborative workshops for SMEs, which will employ gamification techniques to provide engaging privacy training and knowledge dissemination. This initiative will empower participants with the skills to navigate GDPR requirements. Fourth, we will curate comprehensive documentation in the form of a handbook, a holistic resource for those seeking to deepen their understanding of GDPR principles and practices. Finally, to help readers comprehend patterns, we will outline the development process and provide examples of GDPR patterns. This cohesive approach aligns all aspects of our work, from practical tools to educational initiatives, into a unified effort to enhance GDPR compliance.

Pattern Development

Our intention is to increase awareness for requirements related to the EU GDPR. Further targets are practice-oriented knowledge transfer and training. The content side covers the creation of compliant technical implementations and the organizational adaptation. Different GDPR target groups such as data subject, data controller, and data processor are considered. In summary, we define patterns that describe the GDPR requirements

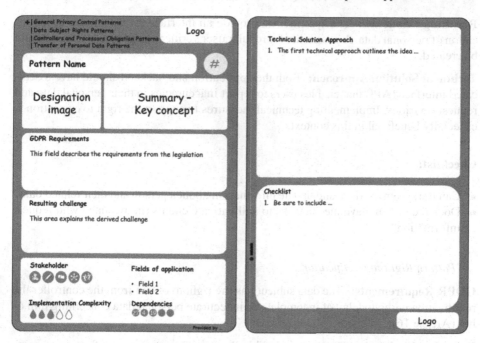

Fig. 3. GDPR Pattern Cards Template.

and provide technical solution strategies for compliance. Our patterns are structured according to a fixed scheme (problem definition, context, solution approach, checklist). To promote the reusability of technical solution approaches, we also reveal relationships between requirements. To do this, we link the patterns with each other accordingly. The patterns serve as blueprints that offer solution strategies for questions that may arise in daily business. GDPR Patterns are aggregated in our pattern catalog as knowledge source.

Privacy Patterns

We introduce four GDPR patterns which extend our current research [6, 7]. The patterns presented here were initially developed in earlier publications, subsequently refined, and are now being published. The *Accuracy pattern* belongs to the General Privacy Control Patterns area. In the Data Subject Rights Patterns area, we add two patterns *Right to Rectification* and *Notification Obligation*. In the area Controllers and Processors, the *Privacy by Default* pattern is introduced (GDPR areas are explained in Sect. 2.2).

Pattern Accuracy

GDPR Requirements: Personal data must be correct and up to date. All reasonable measures must be taken to ensure that personal data, which is inaccurate in relation to the purposes for which it is processed, is deleted or rectified without delay (Article 5 (1d)).

Resulting Challenge: The system must provide an interface for the erasure or rectification of personal data. A possibility for regular user verification of personal data should be created.

Technical Solution Approach: Both the application and backend should have a dedicated interface (API) that enables users to report inaccuracies in their personal data and request revisions. Implementing technical measures to support the right to correction is especially beneficial in this context.

Checklist:

- Can the system retrieve and update information about a person and their related data?
- Does the system have mechanisms to authenticate clients (individuals) who request information?

Pattern Right to Rectification

GDPR Requirements: The data subject has the right to obtain from the controller the rectification without delay of incomplete or inaccurate personal data concerning him or her (Article 16).

Resulting Challenge: The system should allow users to request rectification or completion of personal data. To support this feature, the system needs a flexible architecture that enables data adaptability during operation and should prevent misuse.

Technical Solution Approach: To enable flexible data processing, the backend should provide interfaces for subsequent data processing. This can be achieved using microservices and REST interfaces, offering different options based on needs.

Checklist:

- Does the system support rectification of personal data?
- Can data be changed specifically and separately?
- Is changed data immediately available?

Pattern Notification Obligation

GDPR Requirements: Indicates that the responsible person is obliged to inform all recipients to whom personal data have been disclosed of any rectification, erasure or limitation of the processing of such data. In addition, the data processor is obliged to inform the data subject of these recipients upon request (Article 19).

Resulting Challenge: Any modifications made to personal data must be recorded and transmitted to the data recipient. Authorized recipients and persons to be notified must be known and kept up to date for this purpose.

Technical Solution Approach: A backend process must monitor and update a list of authorized recipients for personal data, providing suitable communication channels for each recipient.

Checklist:

- Is the processing of personal data monitored for changes?
- Is the transfer of data (e.g., updates) logged, including information about the recipient?
- Are data subjects notified in case of any change in processing?

Pattern Privacy by Default

GDPR Requirements: Appropriate technical and organizational measures shall be taken to ensure that the default settings of a service do not patronize users in the collection, processing, storage and disclosure of personal data. This is often referred to as Privacy by Default (Article 25 (2)).

Resulting Challenge: First, it is necessary that the collection, processing, storage, and transfer of data can be technically adjusted to any relevant user context. Only then variable data privacy-friendly default settings are possible.

Technical Solution Approach: We recommend associating data with additional attributes, such as including a purpose attribute with personal data. This allows for attribute-based access control to ensure purpose limitation. When data is provided with appropriate attributes, data protection-friendly characteristics can be defined. For example, the data can be tagged with a generic storage attribute "by default," limiting access to these attributes only for storage purposes, preventing processing. Similarly, a time attribute can be used to set a data's lifetime, restricting further processing after expiration. The data subject should individually set this attribute. These two attributes are just examples, and the service operator should consider these requirements when designing the data model and define suitable attributes along with their specifications.

Checklist:

- Does the system have suitable control attributes that identify the data?
- Can users flexibly adjust settings regarding the processing of personal data?
- Are users not patronized by the system?

5   Conclusion and Outlook

With this article, we seek to encourage GDPR compliance amongst companies, with an emphasis on SMEs. We address three key topics: 1) Identifying key challenges; 2) Reviewing related work; and 3) GPDR catalog system development.

We first focused on the critical obstacles that have impeded the effective enforcement of GDPR regulations for several years since its inception. Several studies and associated research have uncovered potential areas for development in adoption, understanding, training, and awareness. This investigation is motivated by a scientific viewpoint with the goal of understanding the underlying problems and developing workable solutions. This addresses the first research question (RQ1).

Reviewing related work in privacy patterns, we found that several preliminary efforts are available, although they tend to be heterogeneous once considered in the sense of a catalog system. The academic community created 72 privacy patterns while considering different privacy strategies and tactics [27, 31]. The primary focus is on technological execution; however, the demands of SMEs are frequently disregarded, allowing potential for additional research and advancement. To that end, further steps need to be conducted to respond to RQ2.

Finally, we introduced a pattern catalog, which includes specific GDPR patterns. These patterns act as blueprints, offering solution strategies for GDPR-related questions that may arise in daily operations. The catalog can be seen as a toolbox, enhancing existing measures. We intend to extend its contents up to 30 GDPR specific patterns, conveying technical and organizational details in an understandable and user-friendly manner. This concept serves as a roadmap for future activities, such as the indicated development of a pattern system. Additionally, subsequent transfer measures will be developed, continuing from this publication, thus addressing RQ3. The pattern cards template presented above is a contribution to this system.

This study has provided the first three phases of the six-phase design science model, which includes the "build" and "evaluate" fields: the identification and motivation of the problem, the definition of the solution's goals, and the design and development phases [17]. This completes our discussion of the "build" field. The creation and publication of new patterns, as discussed above, is the next step. Up to this point, our approach has been based on theory, primarily focusing on pattern analysis and catalog development. Future research will address the current limitations by focusing on the "evaluate" field (phases four to six). Consequently, all developed patterns will be transformed into method cards, inspired by other research domains such as business model development. These cards (refer to Fig. 3) encapsulate the essential aspects of our designed patterns and deliver relevant knowledge in a user-friendly manner. The method cards will undergo validation in workshop formats involving the target group (SMEs). In this context, we aim to assess the suitability of the method cards for knowledge transfer and training purposes, enhancing the maturity level of the patterns. Ultimately, the pattern catalog will be made publicly accessible, for example, by providing it on an online repository.

In our research, we focus on EU GDPR regulations and aim to expand our pattern catalog with additional patterns from the business context (policies). We also monitor the development of the Digital Services Act (DSA) and incorporate relevant patterns. We consider addressing the challenges posed by AI advancements in data protection by exploring the development of AI privacy patterns. Additionally, the EU's ongoing work on an AI Act presents an opportunity to derive patterns from its requirements.

References

1. Schuster, T., Waidelich, L., Alpers, S.: Datenschutz im Marketing Dialogmarketing Perspektiven 2019/2020, pp. 73–83. Springer Fachmedien Wiesbaden, Wiesbaden (2020). https://doi.org/10.1007/978-3-658-29456-4_5
2. Lindqvist, J.: New challenges to personal data processing agreements: is the GDPR fit to deal with contract, accountability and liability in a world of the Internet of Things? Int. J. Law Inf. Technol. **26**, 45–63 (2018). https://doi.org/10.1093/ijlit/eax024
3. Li, Z.S., Werner, C., Ernst, N., Damian, D.: Towards privacy compliance: a design science study in a small organization. Inf. Softw. Technol. **146**, 106868 (2022). https://doi.org/10.1016/j.infsof.2022.106868

4. GDPR.eu: GDPR Small Business Survey (2019)
5. Buckley, G., Caulfield, T., Becker, I.: "It may be a pain in the backside but..." Insights into the impact of GDPR on business after three years (2021)
6. Rösch, D., et al.: Muster zur praxisorientierten Umsetzung und konformen Nutzung der DSGVO. In: David, K., Geihs, K., Lange, M., Stumme, G. (eds.) INFORMATIK 2019: 50 Jahre Gesellschaft für Informatik – Informatik für Gesellschaft, pp. 297–310, Bonn (2019). https://doi.org/10.18420/inf2019_50
7. Roesch, D., Schuster, T., Waidelich, L., Alpers, S.: Privacy control patterns for compliant application of GDPR. In: AMCIS 2019 Proceedings (2019)
8. Alexander, C., Ishikawa, S., Silverstein, M., Jacobson, M., King, I.F., Angel, S.: A Pattern Language – Towns Buildings Construction. Oxford University Press, Oxford (1977)
9. Fowler, M.: Patterns of Enterprise Application Architecture. Addison-Wesley, Boston (2003)
10. Hoepman, J.-H.: Privacy design strategies. In: Cuppens-Boulahia, N., Cuppens, F., Jajodia, S., Abou El Kalam, A., Sans, T. (eds.) ICT Systems Security and Privacy Protection. IFIP Advances in Information and Communication Technology, vol. 428, pp. 446–459. Springer, Heidelberg (2014). https://doi.org/10.1007/978-3-642-55415-5_38
11. Colesky, M., Caiza, J.C., Del Álamo, J.M., Hoepman, J.-H., Martín, Y.-S.: A system of privacy patterns for user control. In: Haddad, H.M., Wainwright, R.L., Chbeir, R. (eds.) Proceedings of the 33rd Annual ACM Symposium on Applied Computing, pp. 1150–1156. ACM, New York, NY, USA (2018). https://doi.org/10.1145/3167132.3167257
12. Notario, N., et al.: PRIPARE: integrating privacy best practices into a privacy engineering methodology. In: 2015 IEEE Security and Privacy Workshops, pp. 151–158. IEEE (2015). https://doi.org/10.1109/SPW.2015.22
13. Martin, Y.-S., Kung, A.: Methods and tools for GDPR compliance through privacy and data protection engineering. In: 2018 IEEE European Symposium on Security and Privacy Workshops (EuroS&PW), pp. 108–111. IEEE (2018). https://doi.org/10.1109/EuroSPW.2018.00021
14. Regulation (EU) 2016/679 of the European Parliament and of the Council of 27 April 2016 on the protection of natural persons with regard to the processing of personal data and on the free movement of such data, and repealing Directive 95/46/EC (General Data Protection Regulation). Regulation (EU) 2016/679 (2016)
15. Hevner, A., March, S., Park, J., Ram, S.: Design science in information systems research. MIS Q. **28**, 75–105 (2004). https://doi.org/10.2307/25148625
16. Gregor, S., Hevner, A.R.: Positioning and presenting design science research for maximum impact. MIS Q. **37**, 337–355 (2013). https://doi.org/10.25300/MISQ/2013/37.2.01
17. Peffers, K., Tuunanen, T., Rothenberger, M.A., Chatterjee, S.: A design science research methodology for information systems research. J. Manag. Inf. Syst. **24**, 45–77 (2007). https://doi.org/10.2753/MIS0742-1222240302
18. Da Freitas, M.C., Da Mira Silva, M.: GDPR compliance in SMEs: there is much to be done. J. Inf. Syst. Eng. Manage. **3**, 30 (2018). https://doi.org/10.20897/jisem/3941
19. Hirvonen, P., Kari, M.J.: Building situational awareness of GDPR. ECCWS **22**, 575–583 (2023). https://doi.org/10.34190/eccws.22.1.1077
20. Christmann, C., Falkner, J., Horch, A., Kett, H.: Identification of IT security and legal requirements regarding cloud services. In: Lee, Y.W., Becker Westphall, C. (eds.) CLOUD COMPUTING 2015. The Sixth International Conference on Cloud Computing, GRIDs, and Virtualization: 22–27 March 2015, Nice, France. IARIA, Wilmington, DE, USA (2015)
21. Sirur, S., Nurse, J.R.C., Webb, H.: Are we there yet? Understanding the challenges faced in complying with the General Data Protection Regulation (GDPR). arXiv (2018)

22. Pedroso, L.M., Araujo, V.M., Cota, M.P., Paulo Magalhaes, J.: How can GDPR fines help SMEs ensuring the privacy and protection of processed personal data. In: 2021 16th Iberian Conference on Information Systems and Technologies (CISTI), pp. 1–6. IEEE (2021). https://doi.org/10.23919/CISTI52073.2021.9476620

23. Brodin, M.: A framework for GDPR compliance for small- and medium-sized enterprises. Eur. J. Secur. Res. 4(2), 243–264 (2019). https://doi.org/10.1007/s41125-019-00042-z

24. Poritskiy, N., Oliveira, F., Almeida, F.: The benefits and challenges of general data protection regulation for the information technology sector. DPRG 21, 510–524 (2019). https://doi.org/10.1108/DPRG-05-2019-0039

25. Pika, A., Wynn, M.T., Budiono, S., ter Hofstede, A.H.M., van der Aalst, W.M.P., Reijers, H.A.: Privacy-preserving process mining in healthcare. Int. J. Environ. Res. Public Health 17, 1612 (2020). https://doi.org/10.3390/ijerph17051612

26. Huth, D., Matthes, F.: Appropriate technical and organizational measures. identifying privacy engineering approaches to meet GDPR requirements. In: AMCIS 2019 Proceedings, vol. 5 (2019)

27. Colesky, M., et al.: Patterns (2018). https://privacypatterns.org/patterns/

28. Huth, D.: A pattern catalog for GDPR compliant data protection. PoEM Doctoral Consortium, pp. 34–40 (2017)

29. Tikkinen-Piri, C., Rohunen, A., Markkula, J.: EU General Data Protection Regulation: changes and implications for personal data collecting companies. Comput. Law Secur. Rev. 34, 134–153 (2018). https://doi.org/10.1016/j.clsr.2017.05.015

30. Pereira, F., Crocker, P., Leithardt, V.R.: PADRES: tool for PrivAcy Data REgulation and Security. SoftwareX 17, 100895 (2022). https://doi.org/10.1016/j.softx.2021.100895

31. Kargl, F.: Privacy by Design. https://privacybydesign.digital/

An Exploratory Study and Prevention Measures of Mob Lynchings: A Case Study of India

Gautam Kishore Shahi[1][(✉)] and Tim A. Majchrzak[2]

[1] University of Duisburg-Essen, Duisburg, Germany
gautam.shahi@uni-due.de
[2] University of Agder, Kristiansand, Norway

Abstract. Mob lynching is violent human behaviour where people punish someone without a legal trial. Lynching ends with a significant injury or death of a person. In the digital era, almost every piece of news, including the incident of mob lynching, is shared on social media. Once a case of mob lynching is shared on social media, it propagates to a wide range of audiences. This study uses quantitative and qualitative approaches to investigate mob lynching in India and discuss possible prevention measures through social media using two datasets, one collected from Twitter and another from a fact-checking website. The fact-checked data provides an overview of the cause and effect of mob lynching, while the tweets provide the current discussion and user response. The proposed analysis highlights frequent topics discussed on Twitter, user reactions to those posts, and their relation with historic mob lynching incidents. In the end, a prevention measure is proposed using a user study. The result shows that most tweets are of negative sentiment; tweets are more retweeted than likes. The topic modelling shows the tweets are asking for mob lynching and help. There is partial support for prevention measures, and there is a need for literacy, awareness, and strict law to control mob lynching in future.

Keywords: Mob Lynching · Exploratory Analysis · Fact-checked Data · User Reaction · Prevention Measure

1 Introduction

A lynch mob is an angry crowd of people who want to kill someone based on common beliefs or rumours [1]. Social media nowadays play a critical role in provoking mob lynching, which may cause a chain of mob lynchings in different places [2]. Multiple countries are facing the issue of mob lynching. For instance, a case of lynching was reported in India in Patna city for killing a man [3], and another mob lynching incident occurred in Palgarh city [4].

The origin of the word *lynch* is uncertain, but it possibly originated during the American Revolution [1]. Mob lynching happens when an uncontrolled mob "takes justice into their own hands" to punish or harm a suspected accused

© The Author(s), under exclusive license to Springer Nature Switzerland AG 2023
J. Maślankowski et al. (Eds.): PLAIS EuroSymposium 2023, LNBIP 495, pp. 103–118, 2023.
https://doi.org/10.1007/978-3-031-43590-4_7

of a crime instead of trusting the authorities to take proper actions against the accused [5]. The action taken by the mob is without a legal trial. Mob lynching, or the enforcement of (perceived) justice by a mob of people, is a rare phenomenon worldwide [6]. However, there has been an increase in the reported cases of mob lynchings in the last decade, particularly in South Asian countries, for example, in India [7]. India has reported several cases of mob lynchings regarding religious clashes or suspicions of child kidnapping and illegal slaughtering of animals [8]; however, it is a challenge to estimate the number of such lynches.

Different sources state varying numbers of brutal lynchings. The National Crime Record Bureau (NCRB)[1] does not define mob lynching as a separate crime [9]. Hence, no publicly available record will illustrate the number of cases that specifically label a crime as a Mob Lynch. The Indian news website Quint lists 113 mob lynching-related deaths since 2015 [10], whereas Wikipedia lists 46 fatal cases from 2017 to 2018 [11]. Prior researchers mentioned 45 deaths from mob lynchings, but the number of unreported lynchings could be higher [1]. While as per the factcheker.in[2], a fact-checker website in India, from January 2009 to July 2019, there were 301 cases reported in India, which killed 103 people and 523 people got injured. The reported mob lynching incidents include reported cases covered by the news media or police. The dark figure is assumed to be a higher mob lynching incidents than the number of confirmed cases of reported cases by news media [12].

News of mob lynching spread on social media platforms like Twitter, Facebook, and WhatsApp. Mob lynching caused due to the circulation of messages on WhatsApp has even been called *WhatsApp Lynchings* [13]. Due to this, the Indian government blamed the spreading of misinformation on WhatsApp and asked WhatsApp to change some features [14], such as restricting message forwarding to only five contacts. The changes to the WhatsApp forwarding policy did not decrease mob lynching incidents, though [14]. Additionally, the government started restricting Internet service in the incident area when they noticed increasing rumours that could lead to mob lynching.

The spreading of rumours leading to mob lynchings was observed in other social media and news media. In July 2018, the Indian Supreme Court started working on strict laws dealing with mob lynching, but no law or legal framework has been successfully implemented. Social media allows users to share information in real-time and spread it easily [15,16]. There are 27.25 million active users on Twitter in India [17]. On average, each user spends 2.4 h per day on their mobile device [18]. Approximately 18% of social media users in India use Twitter as a news source [19]. Almost all news media organisations are active on Twitter and share the news in real time. This study considers the study of mob lynching on Twitter because it immediately provides data and user interaction. Most cases of mob lynching get reported by the users or news media on Twitter. We propose the following research questions:

---

[1] www.ncrb.gov.in/en.
[2] www.lynch.factchecker.in/.

RQ1: How do users react to the public discourse of mob lynching on social media? To answer the first RQ, we collect Twitter data using hashtags related to mob lynching. To get an overview of the topic discussed in the discourse, topic modelling is used to get the most frequent topic discussed in the tweets. Also, descriptive analysis of tweets is provided to emphasise user reactions in terms of like, and retweets.

RQ2: What preventive measures can be taken to reduce the cases of mob lynchings? Currently, no concrete measures successfully control mob lynching. We have used a user study to gather public support about the understanding and improvement of preventive measures to reduce mob lynching. In the end, we suggest preventive measures to reduce mob lynching.

In this study, we use a mixed method approach; first, an explanatory analysis of fact-checked data is performed to get an overview of incidents that happened in the last few years. We analyse the total number of cases over ten years, the number of cases reported in different states of India, the number of fatalities, the number of cases reported under different ruling parties and the causes of mob lynching. To understand the user interaction with mob lynching on social media, we have analysed the tweets using topic modelling, sentiment analysis and diffusion of tweets using likes and retweets. Since the measure taken by the government was not effective [20], a qualitative approach is added by conducting a user study to ask about the preventive measures.

The remainder of the paper is structured as follows, Sect. 2 discusses the previous research on mob lynching Sect. 3 explains our research design. Then, Sect. 4 shows results and findings, before Sect. 5 provides a discussion. Section 6 provides a conclusion & future work.

2   Related Work

The verb *lynch* means attempting to kill someone who has not been found guilty of a crime at a trial. Several stories describe the origin of the term lynch, one article mentioning the lynch law that implicates the punishment without trial [21]. Extrajudicial punishments happen around the globe [22]. As a result, every society has a history of acts that include violence against people not being proven guilty. However, mob lynching marks a new branch of collective violence. Collective violence is a mob behaviour defined as being spontaneous, sparsely structured, open to change, unstable and short-lived [1]. Furthermore, mob lynching can be considered as an act of *mobocracy*, the domination by the masses [23].

Considering the phenomenon of mob lynching in the bigger picture, its process needs to be examined. First, negative emotions, such as anger, frustration and fear, assemble because of the current situation in which rumours are propagated [21]. Because of these emotions, people want to take the law into their own hands. They are unwilling to wait for administrative action. Consequently, people come together, connect and form a mob to perform an act of punishment that can result in fatality [1,24].

In India, several religious communities live together; Hinduism is the most common religion, with 79.8% of the Indian population being Hindus. 15.38% are

Muslims, 2.44% Christians, and 1.72% are Sikh [25]. Differing beliefs, values and prejudices towards other groups accompany the diversity of religion. This can lead to religious conflicts, such as religiously motivated mob lynchings [26]. One news site reports an average of one religious hate crime happens per week [27]. Mob lynching may be performed due to personal fear when a group of people assume an unresolved threat to them [28]. For instance, people may fear that a child kidnapper is roaming around to abduct a child.

Lynchings triggered by (not proven) accusations occur mainly in rural areas of India. These areas are mostly poor and characterised by low levels of education [24]. The cases can be discovered quickly on Twitter. Using Twitter, the incident spreads faster through retweets and likes, which may trigger another case of mob lynching due to rumours. For example, a newspaper reports, due to rumour on social media, "A woman beggar in Ahmedabad was lynched over suspicion of child theft" [29]. Among other things, these motives need to be examined. Since there is a high diversity of information on mob lynchings, the motives and the phenomenon of mob lynching generally have to be compared across different data sources.

Social media has quickly changed the communication pattern in society. Social media attracts a large audience, and users can easily access it at a low-cost [30,31]. It allows flexible responses like a retweet [32,33], differentiating it from traditional media. Many authors have analysed the role of social media in social crime in society [34–36]. People use social media platforms like WhatsApp, Facebook and Twitter to spread rumours and misinformation [37]. Several incidents of mob lynchings were executed due to the instant messages circulated on the social media group [38]. Rumours over social media may instigate locals to target strangers who can not speak the local language. For instance, rumours of child kidnapping led to a mob lynching in Palghar, India. In this incident, three people, including two seers, were killed by the violent mob [39].

The literature shows that social media plays a crucial role in propagating rumours, which may end with lynching [14]. False information spreads faster than partially false information [40]. Also, the volume of misinformation in India is comparatively high [41]. Rumours act as a catalyst for the people who indulge in lynching activities. Unknowing people become a part of the crime by spreading rumours or misinformation. It is also believed that due to the rumours spread on social media, the cases of mob lynching have increased [42,43].

3   Research Design

For this study, we collected data from two sources: one of the datasets is from a fact-checking website, and the other is from Twitter. The first dataset gives the details of previously reported cases of mob lynching in India; the second dataset provides real-time discussions on Twitter during social unrest.

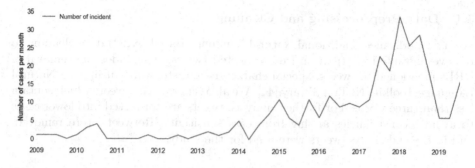

Fig. 1. Timeline of reported Mob Lynching incident (Jan 2009 to mid Jul 2019)

Dataset I – Fact-Checked Data: The fact-checked articles are gathered from the Fact-Check website[3]; the only fact-checking website that compiled the list of mob lynching cases in India from 2009-01-28 to 2019-07-28 (then after website stops providing data). They pick claims and statements from public figures or government reports and check for veracity and context. For every recorded case, the data provided further information on victims, perpetrators, motives, the pretext for the lynchings, the location and date, description, source link, and fact-checked link when the mob lynchings happened. The information on the victims typically includes their religion and the fate they suffered after the mob lynching, for example, if they were injured and if they succumbed to their injuries. Moreover, information is given about the religion of the perpetrator and their punishment if convicted. Overall, 301 reported cases were compiled.

Dataset II – Twitter Data: We collected the dataset from Twitter using a self-developed Python crawler [40] over the Twitter search API. To get user opinions, tweets were collected after fact-checking the data range. When another mob lynching case occurred, tweets were collected from 2019-10-10 to 2020-02-24. The collection was conducted by tracking tweets using the frequently used hashtags in trending tweets, #MobLynching and #lynching and gathered 148,277 tweets. After collecting the data, we noticed tweets from Hong Kong protests, Pakistan, and the USA are included in the dataset. So, to find the hashtags related to mob lynching in India, we further filtered 2,506 hashtags having a frequency of more than 100 times and manually analysed hashtags related to the Indian context from the above list. In the end, we got 55 unique hashtags representing the current incident of mob lynching in India during the above period, such as #lawyersvspolice, #bjp, #beefban, #gangsofcowssey, #hatecrime,#tishazaricourt, #muslims, #rape, #lawyersnotabovelaw, #radicalislamicterrorism,#delhipoliceprotest, #mohanbhagwat, #rapist_jalao, #tabrezansari, #jharkhand #jihadikilling. Overall, we filtered 5,477 tweets, including retweets for this study, which were used for preprocessing.

---

[3] www.factchecker.in.

3.1   Data Preprocessing and Cleaning

We have performed traditional Natural Language Based (NLP) data cleaning to remove unwanted information from collected tweets. It includes the removal of URLs mentioned in tweets, special characters and stopwords using the Natural Language Toolkit (NLTK) library [44]. We also removed the tweets which contain less than three words. For further analysis, tweets are converted into lower cases to avoid inconsistencies arising from case sensitivity. Retweets were removed, and in the end, 1,468 tweets were used for the study.

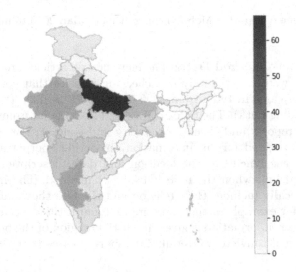

Fig. 2. Location of Mob lynching thought India

We combined the two datasets to emphasise past mob lynching and user interaction to discuss mob lynching on social media. In explanatory analysis, we analyse the number of incidents reported in 10 years. The analysis indicates the different causes of mob lynching and the number of cases for a particular cause over the entire available dataset. We used a topic modelling approach to find the most frequent topics discussed among users and users' responses to tweets regarding likes, retweets, and sentiment analysis.

3.2   Explanatory Analysis

The fact-checked data consists of several pieces of information like date, place of incident, geographical coordinates, fatality, the status of the police report and cause of mob lynching. These attributes present meaningful insights, like how the mob lynching incident was reported across India and the number of reported cases of them and presented on a geographical map. A timeline of a number of mob lynching incidents visualises the actual growth of violent crimes in the country. Some political parties blame each other for supporting incidents of mob lynching, so we also compiled a list of incidents that occur within the ruling party of states. There are multiple causes of mob lynching, and it varies across the different states; a group together of cause and presenting can show the actual status.

Table 1. Descriptive analysis of Twitter Data

Parameter	Value
Number of Tweets	1,486
Unique Account	1,347
Verified Account	14
Tweet without Hashtags	1,486
Tweet without mentions	1,205
Tweet with Emoji	186
Median Retweet Count	95
Median Favourite Count	0
Median Followers Count	344
Median Friends Count	367
Median Account Age (days)	64
Sentiment (positive & negative)	1,064 & 464

3.3 Data Analysis

To answer the first RQ, we used a data analysis technique to analyse the collected data using text and network based features. The detailed description is given in Table 1. Text-based analysis helps finding find meaningful information from texts. For that, we used topic modelling and sentiment analysis.

Topic Modelling is used to find the most important topics discussed in the dataset. The top topics explain the main discussion points in the discourse. We have carried this out by using an unsupervised model, Latent Dirichlet allocation (LDA), for topic modelling [45] and presents the most frequently discussed topics during mob lynching and possibly the causes that might have led to the lynching using a Python program.

Sentiment is an emotion in the textual form. *Sentiment Analysis* is widely used in different social problems. This study uses sentiment as the nature of tweets posted using the hashtags of mob lynching. We used the Python library VADER sentiment tool for sentiment analysis, which provides positive, negative and neutral sentiments [46].

Network-based analysis aids in getting the response of users to tweets related to mob lynching. We have considered likes and retweets as parameters for network interaction. *Likes* indicates a user agrees with the content. Tweets with more likes are more popular and have higher reachability. Retweet means sharing the tweets; retweets are a way of sharing another post. More retweets mean it reaches more users on their timelines [47].

3.4 User Study

To understand the public support regarding mob lynching and measures that need to be taken to reduce it, we conducted a user study as an online survey in India. The questionnaire was divided into different parts intended to focus on the different levels of mob lynching and the research questions associated with them.

The first part of the questionnaire aims to create a common understanding of mob lynching by providing an example and scenario. We then asked the participant what they understood might be the causes of the lynching incident. We also collect a set of demographic data to get their background overview.

The last part of the user study assesses users' opinions on the prevention measures taken by the Government of India to control mob lynching. We evaluate two prevention measures; one is the reduction of the spreading of rumours on WhatsApp by limiting the feature of forwarding messages to only five chats at a time (compared to 250 chats previously) [2], and another is the reduction of rumours related to mob lynching by enforcing the shut down of the internet services in the area of the incident and the surrounding areas [48].

The evaluation includes the indication of support from the participants on three parameters. The first parameter is their opinion on the effectiveness of the measure. The second parameter conforms to the effects of the restrictions on their personal life. Lastly, the third parameter is an open question, where the participants were asked to suggest measures to control mob lynching.

4 Results

This section describes and discusses the results obtained from the above analysis on both datasets and user studies.

4.1 Explanatory Analysis of Fact-Checked Data

A timeline shows the frequency of mob lynchings across India (Fig. 1). The orange color indicates the number of reported mob lynching cases over ten years. The timeline plot of mob lynching shows that the number of reported cases increased until mid-2019. Around 87% of reported cases occurred in the last five years(from 2014 to 2019).

We used the geographical coordinates of mob lynchings to plot the number of incidents reported in each state of India. We used *geopy*, a Python library, to create the geographical map of India and group the number of incidents in each state. We used the colour bar to indicate the number of cases, as shown

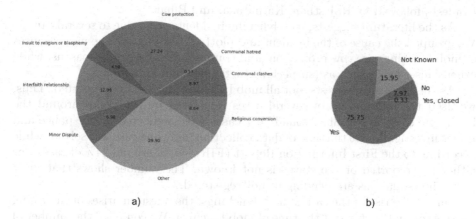

Fig. 3. a) Cause of Mob Lynching in India (in percentage) b) Number of FIR registered for Mob Lynching in India (in percentage)

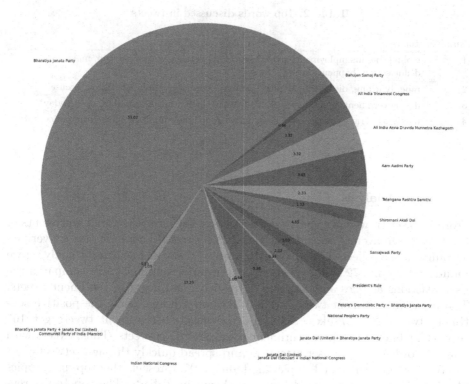

Fig. 4. Mob lynching occurred during the ruling of political parties (in percentage)

in Fig. 2. The number suggests that the maximum cases reported are in Uttar Pradesh, followed by Rajasthan, Karnataka, and Bihar.

As the literature suggests, mob lynching incidents occur due to several causes. We grouped the cause of the incident and plotted it in Fig. 3a. The major causes of mob lynchings are cow protection, and communal and religious reasons, while other causes exist, such as (suspected) theft and child kidnapping.

As the literature suggests, not all mob lynching incidents get reported. Thus, we used the information of several cases registered by the police around the in the pie chart in Fig. 3b. Around 76% cases were registered by the police and under investigation at the time of data collection; only one case was solved, while according to the First Information Report (FIR) [49] of around 24% of cases was either not reported or the status is not known. The number shows that most police investigations are pending or not registered.

Due to the rise in the case of mob lynchings, the question arises of the ruling party's role in the states at the time of mob lynching. We compiled the number of cases that occurred during each ruling party from the data. The distribution of the ruling party at the time of mob lynching is shown in Fig. 4, and the incident of mob lynching was reported under the ruling of several parties.

Table 2. Top words discussed in tweets

Number	Topics
1	moblynche, unemployment, article, lynch, tripletalaq_nr, people, bjp, make, delhielection, happen
2	rape, lynching, die, unemployment, woman, cow, wear, holy_let, tiktok, accuse
3	day, government, indian, go, still, concerned, artist, misinformation, user, today
4	moblynche, rapist_jalao, case, logic, sir_pl_reply, mob, moblynching, stop, police, tal

4.2   Data Analysis

From Dataset II, we have analysed the average likes, retweets, and sentiments of the tweets. On average, each tweet gets 128 retweets. There are only 14 verified accounts in the collected data. We also applied the Vader sentiment analysis; we found that around 72.5% of tweets have a negative sentiment. We compared the user reactions as likes and retweets on positive and negative sentiment tweets. The negative sentiment tweets get 3.2 likes on average, while the positive sentiment tweets get 1.7 likes. Similarly, the negative sentiment tweets get 165 retweets, while the positive sentiment tweets get 31 retweets. Thus, tweets with hateful words get more user attention and spread quickly through retweets.

We further performed topic modelling. We present the top five topics obtained from the topic modelling, as shown in Table 2. The top terms represent the topics discussed related to mob lynching, which indicates triggering mob lynching like, *cow accuse, rapist_jalao* ("burn the rapist") and asking for help like *sir_pl_reply* ("sir please reply"). These discussions are mainly based on tweets related to lynching.

4.3   User Study

We conducted an online user study across India by sharing the survey link across different social media platforms and mailing lists with Indian citizens. There were 33 participants from different parts of India, aged from 22 to 38 years (M = 22.00). 27 out of 33 participants had a university degree; there were 16 females and 17 males. Most of them belonged to the religion of Hinduism (24); the remaining participants identified as Muslims (5), Christians (2), Buddhists (1) and others (1). Table 3 shows the uses of different social media by participants.

Table 3. Use of Social media by participants (in percentage)

	WhatsApp	Facebook	Twitter	Instagram	YouTube
Never	–	–	45.45	33.33	–
Monthly	3.3	3.3	21.21	9.09	3.3
Weekly	–	6.6	3.3	27.27	9.09
Daily	36.36	24.24	18.18	21.21	45.45
Multiple times a day	60.6	66.66	12.12	9.09	42.42

We asked for people's opinions on the government's preventive measures and their suggestions to reduce the cases. For prevention measures, we asked three questions for each possible step. 1) *Do the participants support the measure?* 2) *Do the participants think the measure is effective?* 3) *Does the measure restrict their personal life?*

Qualitative analysis is done to analyse the answers. These answers to the questions are then categorised into the participant's statements, and the reasoning is given optionally. Finally, a short conclusion for each categorisation is discussed below.

Prevention Measure 1. To reduce the spread of rumours, WhatsApp limited the forwarding messages to only five chats at a time (compared with 250 previously).

Response. Participants believe there is no restriction on everyday life but a risk of censorship. Stopping the spread of false information on social media is an important factor; therefore, this measure is mainly supported. Also, the measure is not effective as a stand-alone solution. Education and legislation must be improved.

Prevention Measure 2. To reduce mob lynching, the local government shut down Internet services in the surrounding area to reduce the propagation of rumours about lynchings.

Response. Participants believe it is a controversial measure. They believe that if a restriction occurs, it is a severe restriction of freedom. The balancing of freedom and protection is important. Information spreading is critical, so this measure should only be applied to extreme situations. This measure is not seen as effective in the long term; education and legislation must be improved.

Prevention Measure 3. Besides two measures, the free text field for further suggestions for measures against mob lynching was categorised and listed according to their frequency.

Response. In the open suggestions, 14 participants believe improving the legal system will help in reducing mob lynching. For example, they suggested a "fast and fair judicial delivery systems", "strict laws", to "punish the people", and that "states must take specific preventive, punitive and remedial measures". Around ten participants mentioned raising awareness would decrease the case like "lynching should be seen as a crime", "making people aware of the law and the rights of every person", and "let's start by condemning it". Eight participants support that improving education will also help to reduce the cases like educating the people not to trust everything on social media", "making them think open-mindedly", and "curricula of our education". Four participants believe that decreasing the misinformation will help reduce mob lynching. The administration must also keep an eye on the use of social media. Especially Twitter, WhatsApp groups and forwards and prevent the spread of fear and panic". Three participants believe a change of attitude from government personnel will help in reduction like "police must have the face of a friend and problem solver as opposed to what it is today", "making police do their job, instead of assisting the mob or watching it".

5   Discussion

We have analysed mob lynching as a case study in India by considering the actual incident and discussion on social media. Gathering a large dataset is challenging as the Government of India does not publish records for mob lynching incidents. Not all cases get formally registered, further complicating the situation in analysing the actual situation of mob lynching. Collecting historical or personal data from WhatsApp was not possible, so we used Twitter to analyse user reactions to mob lynching.

The explanatory analysis shows the cause of lynching was mainly religious, as suggested in [8]. There is a sudden rise in the cases of mob lynching [20], and several cases do not get registered.

From the user reaction to discourse on social media, we found that most tweets have negative sentiment, and it propagates faster in terms of retweets, and many people support it in terms of likes compared to tweets with positive sentiments. With more retweets, the diffusion of information gets faster [50], and more people interact with tweets. The topic modelling suggests that tweets show the actual discussion of mob lynching where a group of people ask to conduct mob lynching while others ask for help, so these tweets reflect the actual situation with fact-checked data.

For the second research question, we concluded that most of the participants supported the current platform restriction measures while opposing the internet shutdown as it indicates violations of people's personal life.

For preventive measures, people want to create awareness about mob lynching, reduce misinformation, enhance legislation for the perpetrators, and improve police work.

While we have been able to gain insights that foster a better understanding of the relationship between mob lynching and the use of social media, our study has limitations. First, we were bound to data from a fact-checking Web site due to the lack of official data from NCRB. This poses a potential bias, meaning that essential data might have been omitted. However, currently, this is the most promising data source available. Second, our qualitative study was very limited in scope due to the small number of participants. It fits with the exploratory nature of our work, but all insights gained are anecdotal, but at least part of the entire mob lynching incidents. Still, what we learned from the participants forms the foundation for describing which steps should be taken next. Third, no data could be collected from WhatsApp. This limitation concerns all kinds of messaging services, particularly any communication through encrypted channels. Fourthly, due to the prevalent nature of the word *lynching*, the Twitter data we collected was quite noisy. There is no proper scientific publication for many claims, so we rely on news media as a source of information.

6   Conclusion and Future Work

In this work, mob lynching, an ongoing issue in India, was analysed. We used both social media data and factual data. The results show no common trends in the incidents of mob lynching across different states of India. Most cases are communal based, and many reported failure to be controlled by state authorities. Several reported cases are still under investigation. User interaction with different tweets with many likes and retweets could be a reason for the fast diffusion of news of mob lynching. We have formulated public support from the user study regarding preventing mob lynching. Apart from the current preventive measures from the government, users also suggested their opinion on controlling mob lynching.

A continuation of this work could be collecting data from multiple social media platforms such as public WhatsApp groups, Reddit, and Youtube to get more detailed views from user-generated content. For the user study, we could increase the number of participants for a more generalised public opinion. Another potential direction is analysing the propagation of misinformation during a mob lynching on encrypted (like WhatsApp) [51] and non-encrypted platforms (like Twitter, YouTube) [52] and how users respond to discourse. The case of mob lynchings has increased in the last five years (2014–2019), which is aligned with the growth of internet users and misinformation. Another possible direction of future work would be to analyse the correlations and causations among the role of internet users, social media, and misinformation triggering mob lynching.

References

1. Jha, R.S., Jain, V., Chawla, C.: Hate speech & mob lynching: a study of its relations, impacts & regulating laws. Think India (Q. J.) **22**(3), 1401–1405 (2019)
2. Mukherjee, R.: Mobile witnessing on Whatsapp: vigilante virality and the anatomy of mob lynching. South Asian Popular Cult. **18**(1), 79–101 (2020)
3. Akhef, M.: Patna: Man lynched for killing 11-year-old punpun boy (2023). https://timesofindia.indiatimes.com/city/patna/patna-man-lynched-for-killing-11-year-old-punpun-boy/articleshow/99179725.cms
4. Wikipedia. Mob 'lynching of Arab' aired live on Israeli TV (2020). https://en.wikipedia.org/wiki/2020_Palghar_mob_lynching
5. Garland, D.: Penal excess and surplus meaning: public torture lynchings in twentieth-century America. Law Soc. Rev. **39**(4), 793–834 (2005)
6. Blocker Jr., J.S., et al.: Lynching Beyond Dixie: American Mob Violence Outside the South. University of Illinois Press, Urbana-Champaign (2013)
7. Bhat, M., Bajaj, V., Kumar, S.A.: The crime vanishes: mob lynching, hate crime, and police discretion in India. Jindal Global Law Rev. **11**(1), 33–59 (2020)
8. Janghu, P., Pranjal, B.: Emerging trend of violent cow protection and the right to religion in India
9. Tripathi, K.: No reliable data to define mob-lynching as a separate category of crime (2022). https://www.financialexpress.com/india-news/ncrb-no-reliable-data-to-define-mob-lynching-as-a-separate-category-of-crime-modi-govt-nityanand-rai/1787225
10. Quint. Lynching in India (2022). https://www.thequint.com/quintlab/lynching-in-india/
11. Gupta, I.: Mob violence and vigilantism in India. World Affairs J. Int. Issues **23**(4), 152–172 (2019)
12. Baksi, S., Nagarajan, A.: Mob lynchings in India: a look at data and the story behind the numbers.Newslaundry.com, 4 July 2017
13. Vasudeva, F., Barkdull, N.: Whatsapp in India? A case study of social media related lynchings. Soc. Identities **26**(5), 574–589 (2020)
14. Arun, C.: On whatsapp, rumours, lynchings, and the Indian government. Econ. Polit. Weekly **54**(6), 30–35 (2019)
15. Shahi, G.K., Tsoplefack, W.K.: Mitigating harmful content on social media using an interactive user interface. In: Hopfgartner, F., Jaidka, K., Mayr, P., Jose, J., Breitsohl, J. (eds.) Social Informatics: 13th International Conference, SocInfo 2022, Glasgow, UK, 19–21 October 2022, Proceedings, vol. 13618, pp. 490–505. Springer, Cham (2022). https://doi.org/10.1007/978-3-031-19097-1_34
16. Nandini, D., Schmid, U.: Explaining hate speech classification with model agnostic methods. arXiv preprint arXiv:2306.00021 (2023)
17. Keap, S.: DIGITAL 2023: INDIA (2023). https://datareportal.com/reports/digital-2023-india
18. Knowbles, T.: Two and a half hours a day spent staring at our smartphone (2022). https://www.thetimes.co.uk/article/two-and-a-half-hours-a-day-spent-staring-at-our-smartphones-llvlp3t3v

19. Sannam S4. Top social media trends in India (2020). https://acumen.education/top-social-media-trends-in-india/
20. Kohli, I.: Mob lynching in India: is the government doing enough? Available at SSRN 4341681 (2023)
21. Satonkar, S.D.: Mob lynching: collectively harmful influences of social media and hidden irrational thoughts of REBT. Phoenix Int. J. Psychol. Soc. Sci. **3**, 120–128 (2019)
22. Carrigan, W.D.: Lynching Reconsidered: New Perspectives in the Study of Mob Violence. Routledge, London (2014)
23. Ellsworth, P.D.: Mobocracy and the rule of law: American press reaction to the murder of Joseph Smith. Brigham Young Univ. Stud. **20**(1), 71–82 (1979)
24. Choubey, M.: Mob lynching: a case study Jharkhand. Jamshedpur Res. Rev. **4**, 16–21 (2019)
25. O'Neill, A.: India: Religious affiliation in 2011 (2023). https://www.statista.com/statistics/261308/religious-affiliation-in-india/
26. Barz, H.: Islam und bildung: Bemerkungen zu einem ambivalenten verhältnis. Islam und Bildung, pp. 241–262 (2018)
27. Welle, D.: India Struggles with Religious Lynchings (2022). https://www.dw.com/en/india-struggles-with-religious-lynchings/a-49950223
28. Singay, K.A., Baig, T., Ahmad, M.A.: Mob lynching in Pakistan: an integrated conceptual model. Pak. Soc. Sci. Rev. **4**(1), 688–698 (2020)
29. Pennews: A woman beggar in Ahmedabad was lynched over suspicion of child theft (2018). https://www.newsclick.in/mob-lynching-over-social-media-rumours-continue
30. Bekkers, V., Moody, R., Edwards, A.: Micro-mobilization, social media and coping strategies: some Dutch experiences. Policy Internet **3**(4), 1–29 (2011)
31. Mergel, I.: Open collaboration in the public sector: the case of social coding on GitHub. Gov. Inf. Q. **32**(4), 464–472 (2015)
32. Thacher, D., Rein, M.: Managing value conflict in public policy. Governance **17**(4), 457–486 (2004)
33. Ferreyra, N.E.D., Shahi, G.K., Tony, C., Stieglitz, S., Scandariato, R.: Regret, delete, (do not) repeat: an analysis of self-cleaning practices on Twitter after the outbreak of the Covid-19 pandemic. In: Extended Abstracts of the 2023 CHI Conference on Human Factors in Computing Systems, pp. 1–7 (2023)
34. Beshears, M.L.: Effectiveness of police social media use. Am. J. Crim. Justice **42**(3), 489–501 (2017)
35. Meijer, A.J., Torenvlied, R.: Social media and the new organization of government communications: an empirical analysis of Twitter usage by the Dutch police. Am. Rev. Public Adm. **46**(2), 143–161 (2016)
36. Davis, E.F., Alves, A.A., Sklansky, D.A.: Social media and police leadership: lessons from Boston. Australas. Policing **6**(1), 10–16 (2014)
37. Shahi, G.K., Majchrzak, T.A.: AMUSED: an annotation framework of multimodal social media data. In: Sanfilippo, F., Granmo, O.C., Yayilgan, S.Y., Bajwa, I.S. (eds.) Intelligent Technologies and Applications, INTAP 2021. CCIS, vol. 1616, pp. 287–299. Springer, Cham (2021). https://doi.org/10.1007/978-3-031-10525-8_23
38. Gowen, A.: As Mob Lynchings Fueled by WhatsApp Sweep India, Authorities Struggle to Combat Fake News, pp. NA-NA. The Washington Post (2018)
39. Singh, V.: Mob lynching: an expression threatening the right to life. Available at SSRN 3604289 (2020)
40. Shahi, G.K., Dirkson, A., Majchrzak, T.A.: An exploratory study of Covid-19 misinformation on Twitter. Online Soc. Netw. Media **22**, 100104 (2021)

41. Shahi, G.K., Nandini, D.: FakeCovid-a multilingual cross-domain fact check news dataset for Covid-19. arXiv preprint arXiv:2006.11343 (2020)
42. Citizens Against Hate. Lynching without end: report of fact finding into religiously motivated vigilante violence in India, New Delhi, 9 January 2017
43. Sharma, R.: Lynchings in the time of voyeurism and social media misuse (2022). https://economictimes.indiatimes.com/news/politics-and-nation/view-lynchings-in-the-time-of-voyeurism-and-social-media-misuse/articleshow/65132202.cms
44. Loper, E., Bird, S.: NLTK: the natural language toolkit. arXiv preprint cs/0205028 (2002)
45. Blei, D.M., Ng, A.Y., Jordan, M.I.: Latent Dirichlet allocation. J. Mach. Learn. Res. **3**(Jan), 993–1022 (2003)
46. Hutto, C., Gilbert, E.: VADER: a parsimonious rule-based model for sentiment analysis of social media text. In: Proceedings of the International AAAI Conference on Web and Social Media, vol. 8, pp. 216–225 (2014)
47. Shahi, G.K., Clausen, S., Stieglitz, S.: Who shapes crisis communication on Twitter? An analysis of German influencers during the Covid-19 pandemic. In: Proceedings of the 55th Hawaii International Conference on System Sciences (2022)
48. Srivastava, P.: Internet blocked to nip lynch backlash in Meerut (2019). https://www.telegraphindia.com/india/internet-blocked-to-nip-lynch-backlash-in-meerut/cid/1693615
49. Professor of Law. FIR and Role of Police: Legislative and Judicial Trends. Regal Publications (2017)
50. Stieglitz, S., Dang-Xuan, L.: Emotions and information diffusion in social media-sentiment of microblogs and sharing behavior. J. Manag. Inf. Syst. **29**(4), 217–248 (2013)
51. Kazemi, A., Garimella, K., Shahi, G.K., Gaffney, D., Hale, S.A.: Research note: tiplines to uncover misinformation on encrypted platforms: a case study of the 2019 Indian general election on WhatsApp. Harvard Kennedy School Misinformation Review (2022)
52. Röchert, D., Shahi, G.K., Neubaum, G., Ross, B., Stieglitz, S.: The networked context of Covid-19 misinformation: informational homogeneity on Youtube at the beginning of the pandemic. Online Soc. Netw. Media **26**, 100164 (2021)

Author Index

J. Maślankowski et al. (Eds.): PLAIS EuroSymposium 2023, LNBIP 495, p. 119, 2023.
https://doi.org/10.1007/978-3-031-43590-4

Printed in the United States
by Baker & Taylor Publisher Services